T0280321

Modern Birkhäuser Classics

Many of the original research and survey monographs, as well as textbooks, in pure and applied mathematics published by Birkhäuser in recent decades have been groundbreaking and have come to be regarded as foundational to the subject. Through the MBC Series, a select number of these modern classics, entirely uncorrected, are being re-released in paperback (and as eBooks) to ensure that these treasures remain accessible to new generations of students, scholars, and researchers

Hidenori Kimura

Chain-Scattering Approach to H^∞ Control

Reprint of the 1997 Edition

Hidenori Kimura
Department of Mathematical Engineering and Information Physics
The University of Tokyo
Tokyo, Japan

Originally published in the series *Systems & Control: Foundations & Applications*

ISBN 978-0-8176-8330-6 ISBN 978-0-8176-8331-3 (eBook)
DOI 10.1007/978-0-8176-8331-3
Springer New York Heidelberg Dordrecht London

Printed on acid-free paper

www.birkhauser-science.com

Hidenori Kimura

Chain-Scattering Approach to H^∞ Control

Birkhäuser
Boston • Basel • Berlin

Hidenori Kimura
Department of Mathematical Engineering
and Information Physics
The University of Tokyo
Tokyo, Japan

Library of Congress Cataloging In-Publication Data

Kimura, Hidenori, 1941-
Chain-scattering approach to H-infinity control / Hidenori Kimura.
p. cm. -- (Systems & control)
ISBN 0-8176-3787-7 -- ISBN 3-7643-3787-7
1. H [infinity symbol] control. I. Title. II. Series.
QA402.3.K465 1996 95-48746
629.8'312--dc21 CIP

Printed on acid-free paper

ISBN 0-8176-3787-7
ISBN 3-7643-3787-7

Reformatted from author's disk in LATEX by Texniques, Inc., Boston, MA
Printed and bound by Quinn-Woodbine, Woodbine, NJ

9 8 7 6 5 4 3 2 1

PREFACE

Through its rapid progress in the last decade, H^∞control became an established control technology to achieve desirable performances of control systems. Several highly developed software packages are now available to easily compute an H^∞controller for anybody who wishes to use H^∞control.

It is questionable, however, that theoretical implications of H^∞control are well understood by the majority of its users. It is true that H^∞control theory is harder to learn due to its intrinsic mathematical nature, and it may not be necessary for those who simply want to apply it to understand the whole body of the theory. In general, however, the more we understand the theory, the better we can use it. It is at least helpful for selecting the design options in reasonable ways to know the theoretical core of H^∞control.

The question arises: *What is the theoretical core of H^∞control?* I wonder whether the majority of control theorists can answer this question with confidence. Some theorists may say that the interpolation theory is the true essence of H^∞control, whereas others may assert that unitary dilation is the fundamental underlying idea of H^∞control. The J-spectral factorization is also well known as a framework of H^∞control. A substantial number of researchers may take differential game as the most salient feature of H^∞control, and others may assert that the Bounded Real Lemma is the most fundamental building block. It may be argued that, since the Bounded Real Lemma is just a special case of linear matrix inequality (LMI), the LMI is more fundamental as the theoretical foundation of H^∞control and is a panacea that eliminates all the difficulties of H^∞control.

All these opinions contain some truth. It is remarkable that H^∞control allows such a multitude of approaches. It looks entirely different from different viewpoints. This fact certainly implies that H^∞control theory is quite rich in logical structure and is versatile as an engineering tool. However, the original question of what is the theoretical core of H^∞control remains unanswered. Indeed, every fundamental notion mentioned has a method of solving the H^∞control problem associated with it. Unfortunately, however, lengthy chains of reasoning and highly technical manipulations are their common characteristic features. For instance, the method of unitary dilation, which first gave the complete solution to the H^∞control problem, requires several steps of problem reductions starting with the *Nehari problem* [28]. The game-theoretic approach seems to be the most comprehensible because it reduces the problem to a simple completion of squares. This approach, however, introduces unnecessary complications concerning the initial condition. The issue of internal stability as well as the optimality is not well addressed by this approach. It is indeed remarkable that we have not yet found a proper framework to describe H^∞control theory in a clear and self-contained way so that the intrinsic features are explicitly exposed.

This book is a compilation of the author's recent work as an attempt to give a unified, systematic, and self-contained exposition of H^∞control theory based on the three key notions: *chain-scattering representation, conjugation*, and *J-lossless factorization.* In this new framework, the H^∞control problem is reduced to a J-lossless factorization of chain-scattering representation of the plant. This is comparable with the similar fact that the LQG (Linear Quadratic Gaussian) problem is reduced to the Wiener-Hopf factorization of a transfer function associated with the plant and the performance index. It is the author's belief that the approach proposed in this book is the simplest way to expose the fundamental structure of H^∞control embodying all its essential features of H^∞control that are relevant to the design of control systems.

Our scenario of solving the problem is roughly described as follows.

(1) The cascade structure of feedback systems is exploited based on the chain-scattering representation, and H^∞control is embedded in the more general framework of cascade synthesis.

(2) The H^∞control problem is reduced to a new type of factorization problem called a J-lossless factorization, within the cascade framework of control system design.

(3) The factorization problem is solved based on J-lossless conjugation.

Our approach is not entirely new. Among the three key notions, the first two are already well known. The chain-scattering representation is used extensively in various fields of engineering to represent the scattering properties of the physical system, although the term *chain-scattering* is not widely used except in circuit theory and signal processing where this notion plays a fundamental role. It was introduced in the control literature by the author. J-lossless factorization is an alternative expression of the well-known factorization in the literature of H^∞control which is usually referred to as *J-spectral factorization* in order to facilitate the cascade structure of synthesis. The conjugation is entirely new and was formulated by the author in 1989 as a state-space representation of the classical interpolation theory. It turned out to be a powerful tool for carrying out the J-lossless factorization.

Finally, the approach in this book throws a new light on the meaning of duality in H^∞control. In the chain-scattering formalism, *the dual is equivalent to the inverse.* The duality between the well-known two Riccati equations is explained as a natural consequence of the chain-scattering formalism itself.

In writing the book, we considered the following points to be of paramount importance.

(1) The exposition must be self-contained.
(2) The technicalities should be reduced to a minimum.
(3) The theory must be accessible to engineers.

Concerning point (1), the reader is not required to have a background of linear system theory beyond the very elementary level which is actually supplied in this book in Chapters 2 and 3. All the technical complications are due to the augmentations of the plant in forming the chain-scattering representation. These augmentations have their own relevance to the synthesis problem beyond the technicalities. In Chapters 2 and 3, some basic preliminaries of linear system theory and associated mathematical results are briefly reviewed. Chapter 4 is devoted to the introduction of chain-scattering representations of the plant. Various properties of the chain-scattering representations are discussed. The J-lossless matrix is introduced in this chapter as a chain-scattering representation of lossless systems. The notion of J-lossless conjugation is introduced in Chapter 5 with emphasis on its relation to classical interpolation problems. Chapter 6 deals with J-lossless factorization based on J-lossless conjugation. The well-known Riccati equations are introduced characterizing the existence of J-lossless factorization. The H^∞control problem is formulated in Chapter 7 in the frequency domain. The problem is reduced to the J-lossless factorization of the chain-scattering representation of the plant. The state-space solution to the H^∞control problem is obtained in Chapter 8. Chapter 9 discusses the closed-loop structure of H^∞control systems.

This book was projected five years ago, but the writing didn't go smoothly due to the author's lack of time to concentrate on it. The first drive to write the book was obtained when the author stayed at Delft University of Technology for three months in the summer of 1994. Professor Dewilde taught me a lot about the deep theory of chain-scattering representation through stimulating discussions. I would like to express my sincere gratitude to him for giving me this unusual opportunity. The second drive to complete the book was gained when the author stayed at UC Berkeley as a Springer Professor in the Spring of 1995. I was given the good fortune to deliver a series of lectures on H^∞ control. This book took its final shape during these lectures. I owe a great deal to Professor Tomizuka for giving me this valuable experience and to Professor Packard for his patience in attending all of my lectures. My gratitude also goes to my colleagues, Dr. Kawatani, Professor Fujii, Dr. Hashimoto, Dr. Ohta, and Dr. Yamamoto, and many students who helped me in my research on H^∞control; Mr. Okunishi, Dr. Xin, Dr. Zhou, Dr. Baramov, Miss Kongprawechnon, Mr. Ushida, and many others. I am greatly indebted to a research group on control theory in Japan which gave me continual encouragement during the last decade through the private workshops, colloqia, correspondences, and so on. The discussions with Professors Hosoe, Mita, and Hara gave me many ideas and suggestions. I am also grateful to Messrs. Monden, Oishi, Suh, and Oku, and Mrs. Ushida who typed my manuscript and made extensive corrections through the careful reading of the draft. Finally, I am thankful to Ms. Yoshiko Kitagami who created nice artwork for the cover of this book.

Contents

Chapter 1

Introduction

1.1 Impacts of H^∞Control

Before 1960, control theory was composed of several classical theorems such as the Routh-Hurwitz stability criterion, Nyquist stability theorem, Bode's dispersion relations, Wiener's realizability criterion and factorization theory on the one hand, and a set of design algorithms such as lead-lag compensation, Smith's prediction method, Ziegler-Nichols ultimate sensitivity method, Evans' root locus method, and the like on the other hand. Although the preceding theoretical results were highly respected, they were not closely related to each other enough to form a systematic theory of control. However, classical design algorithms such as lead-lag compensation were not clearly formalized as a design problem to be solved, but rather as a tool of practices largely dependent on the cut-and-try process. They were usually explained through a set of examples. In classical textbooks of control theory, we frequently encounter statements that emphasize the cut-and-try nature of control system design such as the following one:

In general, however, a certain amount of trial-and-error analysis is necessary, especially if the specifications are not reading phrased simply in terms of unknown parameters [94, p.335].

The problem is whether the cut-and-try nature is the intrinsic attribute of control system design or if it is due to the incompleteness of our theory. We have no complete answer to this problem for the moment, but it is clear that in the classical era, the cut-and-try was regarded as an important ingredient of control system design. In summary, classical control theory was composed of two parts: several great theorems and highly heuristic design methods. These two parts are coupled only loosely and indirectly.

The advent of the modern state-space approach in 1960 owing to Kalman [46][45] innovated the fundamental structure of control theory. Several independent classical theorems were replaced by a logical stream represented by the definition-theorem-proof format. Design procedures were stated as the solutions of control problems formulated in terms of the state-space description of the plant, rather than cut-and-try-based practices. As modern control theory successfully advanced, a new research field called *dynamical system theory* emerged which covered not only control theory but also circuit theory, signal processing, and estimation theory. Through far-reaching progress of linear system theory in the 1960s, control theory evolved into a research field with abundant highly-sophisticated theoretical topics from those of eaborating cut-and-try practices.

We must notice, however, that modern control theory was created not from practical needs in industry, but from purely academic motivation [46]. Through its rapid progress in the 1960s, it had little impact on control practices in industry, except for the space and military industries. It became a target of criticism by control engineers working in industry and classical control theorists from older generations. They claimed that modern control theory wasn't relevant to real world control problems due to its unrealistic and over-simplified problem formulation. Theory/practice gap was a popular issue of arguments inside the control community.

The arguments on this issue sometimes triggered an emotional antagonization between the two camps. Those who were with modern control theory were forced to prove that the modern approach was suitable for the design of contemporary control systems. It was true, however, that 99% of control systems working in industry were designed *untheoretically* at that time. The sharp difference between the elegance of modern control theory and the complexity of real systems was not only clear but also overwhelming to many people. Those who were reluctant to recognize applicability of modern theory tended to respect classical control theory in which theoretical results were only loosely coupled with practical designs which are highly *ad hoc*.

The arguments of modern versus classical control theories sometimes led to the debate on the superiority between the time-domain and the frequency-domain frameworks. Those who sided with the time-domain analysis and synthesis claimed that the time-domain (state-space) methods were suitable for large-scale computations, whereas those who had affection for frequency-domain treatments claimed the clarity and tractability of the design philosophy based on transfer functions [42]. Those arguments focused upon dichotomies such as

Modern versus Classical

Time-Domain versus Frequency-Domain

Theory versus Practice

which were embedded in the development of control theory during the 1960s and 1970s. Although these dichotomies made the control community unnecessarily argumentative, they gave strong incentives for the further progress of control theory. The existence of these dichotomies was clear evidence that our theory was far from complete.

There have been repeated attempts to establish a new paradigm of control theory in order to reconcile these dichotomies. A typical example was the frequency-domain design methods of multivariable control systems proposed by the British School [83][73][74]. An older but still continuing effort along this line was the *quantitative feedback theory* by Horowitz and his colleagues [40][41]. Many of these attempts tended to de-emphasize the role of theory in the design of control systems, because classical theory never formulated the design problem in a rigorous way and left the design in the realm of experience and insight. For instance, the *Inverse Nyquist Array Method* proposed by Rosenbrock [83] seemed to be a sensible way to carry out the design. But theoretically, it was too naive. It seemed disconnected from the great theoretical contributions of Rosenbrock himself to system theory based on the singular pencil [82].

What was really required to reconcile the dichotomies we were faced with was to formulate the design problems classical theory dealt with in a clear and rigorous way. Zames seemed to be the first person who was aware of this need and who actually tried to do this job [107].

H^∞control theory was the first successful attempt to reconcile ("aufheben") various dichotomies by *deepening*, not by de-emphasizing, the role of theory. H^∞control theory has almost completely eliminated the dichotomy of modern versus classical control theories by formulating the design issues of classical control properly and has solved it based on the state-space tools of modern theory. Classical methods of frequency-domain synthesis have revived in a more general, rigorous and elaborate context supported by powerful state-space tools. The theory-practice gap was no longer a significant issue at the beginning of the 1990s due to a number of successful applications of H^∞control to real design problems, especially applications of H^∞ based robust control theory.

Since the beginning of the 1980s, it was gradually recognized that the real issue of control engineering we were faced with was the difficulty of modeling the plant to be controlled. Robust control theory became the most popular area that was expected to deal with model uncertainty [84]. H^∞control timely supplied a powerful tool for robust control. Summing up, the dichotomies created by the birth of modern control theory that characterized the control engineering of the 1960s and 1970s were almost eliminated by the advent of H^∞-based robust control theory. It should be noted that there were some other important factors which made these dichotomies obsolete. We should not underestimate the impact of the rapid progress of microelectronics which has contributed to economically develop and implement the sophisticated algorithms of modern control theory.

The increasing demand for higher performance of control cannot be satisfied by naive tuning of controllers. As the need grows to control more complex systems, the difficulty of modeling manifests itself. This is a crucial defect of traditional model-based control strategy including H^∞control. Being faced with the difficulty of modeling plant dynamics, methods of model-free control such as fuzzy control and rule-based control began to receive a considerable amount of attention from practical control engineers. These new types of control strategies have created a new dichotomy of

Model-based versus Model-free,

which is much more crucial and fundamental than those of the 1960s and 1970s. A detailed discussion of this dichotomy is beyond the scope of this book.

Now we discuss several distinguished features of H^∞control to concretize the preceding descriptions. The block diagram of Figure 1.1 is now used extensively in the control literature as a natural scheme for describing the plant and the controller. Actually, the convention of taking z as the *controlled error* rather than the controlled variable in Figure 1.1 is relatively new and was brought to the control literature by H^∞control. It already appeared in 1984 in Doyle [22]. The scheme of Figure 1.1 is versatile in expressing various performance indices assigned to closed-loop systems including those for *LQG* in the circumstance that trade offs among various design specifications are required. Now the scheme of Figure 1.1 has been evolved to that of Figure 1.2, to describe the plant uncertainty in a more explicit way. The most advanced robust control strategies evolve around Figure 1.2 in which H^∞control plays a crucial role. Perhaps the largest impact of H^∞control is its ability to deal with plant uncertainties in a proper way. This ability of H^∞control is due to the fact that most robustness problems are reduced to the *small gain theorem* and the H^∞norm, or its extended form is the only relevant quantitative measure for this theorem [106]. Robust control theory was already an active research area in the early 1980s. The central tool there was the singular value decomposition [62][25]. The notion of singular value is very close to H^∞norm, and it was just a matter of time until the two fields merged.

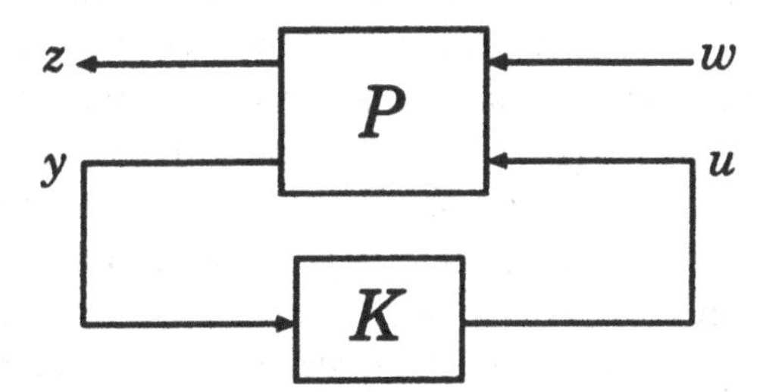

Figure 1.1 Description of Control Systems.

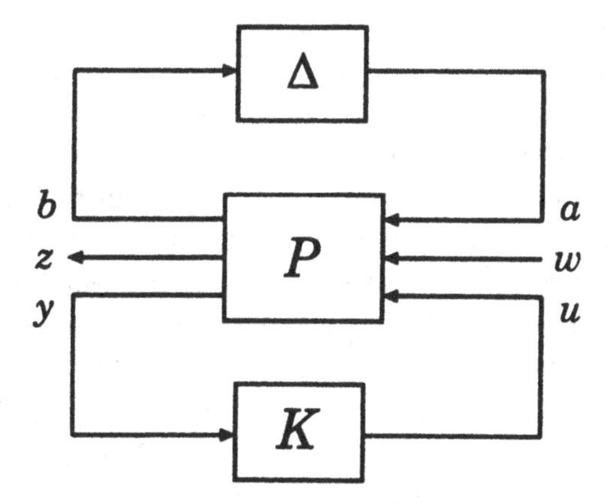

Figure 1.2 Description of Control Systems with Uncertainties.

Actually, things didn't go so directly. H^∞control strategy was first connected intrinsically with a synthesis aspect of robust control in [50] implicitly. The connection was made explicit in [32], and the robust stabilization problem became the simplest case of H^∞control. Now H^∞control is a powerful and indispensable tool for more sophisticated design methods of robust control such as μ-synthesis [21].

In addition to the ability of dealing with various robustness issues in the synthesis of control systems, H^∞control can be used to achieve a desired shaping of the loop transfer function, as is discussed in [75]. Choosing appropriate frequency-dependent weights is a crucial task rather than choosing controller parameters as in the classical approach. Once a set of weights is chosen, the computation enters in the modern framework of state-space. Since the shaping of the loop transfer function is a typical method of classical control system designs such as lead-lag compensation, we can say that H^∞control has successfully amalgamated classical and modern control methodologies, as was addressed previously. This is indeed a novel contribution of H^∞control theory. The shaping of the loop transfer function, however, is effective only approximately, although it is simple and clear [25]. The true power of H^∞control is demonstrated as a method of closed-loop shaping in the frequency domain under the need for tradeoffs among various constraints and specifications. Since the closed-loop shaping is more directly connected to design objectives and is clearly done by H^∞control, we no longer need to rely on open-loop shaping [75].

Looking into the structure of an H^∞controller in detail, we find that it is nothing but a familiar *quasi-state feedback* with observer just like an *LQG* controller. However, there is an important structural difference between H^∞ and *LQG* controllers. The H^∞controller contains the estimated worst-case exogenous signal. The worst-case point of view is another salient feature of H^∞control which distinguishes H^∞control from others. This nature of H^∞controllers comes from the fact that the H^∞ norm is an induced norm. The controller assumes that the exogenous signal (disturbance and/or noise) enters in the worst way for the controller as if it were manipulated by a malicious *anti-controller*. Extensive analysis of this feature of H^∞control was made in [9][70].

Finally, it is worth noting that H^∞control includes *LQG* control as a limiting case. More precisely, as the specified closed-loop norm bound (normally denoted by γ) tends to infinity, the so-called central solution of the H^∞control problem converges to the corresponding *LQG* control.

The situation can be compared to the fact that Newtonian mechanics is an extreme case of quantum mechanics when the Planck constant goes to zero. Therefore, the question, *which is superior, LQG or H^∞*, is nonsense. The H^∞control is at a higher level of logical state than LQG control.

1.2 Theoretical Background

In the early stage of the development of H^∞control theory, the classical interpolation theory of Nevanlinna and Pick [77][78] played a fundamental role. Before this theory was brought forward by Francis and Zames in [29] in an advanced form, some optimization techniques in general Banach space were used in [107][108] that seemed to be much weaker than the familiar optimization method by orthogonal projection in Hilbert space which is used extensively in LQG theory. The interpolation theory was used in a more elementary way for solving the disturbance reduction problem [12] and the robust stabilization problem [50] independently. The application of interpolation was really the first thrust towards the complete solution.

Let us just briefly explain the relevance of the interpolation problem to H^∞control. In Figure 1.1, we assume that the plant $P(s)$ is described as

$$\begin{bmatrix} z(s) \\ y(s) \end{bmatrix} = P(s)\begin{bmatrix} w(s) \\ u(s) \end{bmatrix} = \begin{bmatrix} P_{11}(s) & P_{12}(s) \\ P_{21}(s) & P_{22}(s) \end{bmatrix}\begin{bmatrix} w(s) \\ u(s) \end{bmatrix}. \tag{1.1}$$

Then the closed-loop transfer function $\Phi(s)$ from w to z is given by

$$\Phi(s) = P_{11}(s) + P_{12}(s)K(s)(I - P_{22}(s)K(s))^{-1}P_{21}(s). \tag{1.2}$$

Let us consider first the case where $P_{ij}(s)$ $(i = 1,2;\ j = 1,2)$ are all scalar. Assume that $P_{12}(s)$ has a zero in the right half plane, that is, $P_{12}(z_1) = 0$, Re $[z_1] \geq 0$. The requirement of internal stability prohibits the cancellation of this unstable zero by a pole of $K(s)(I - P_{22}(s)K(s))^{-1}P_{21}(s)$ in (1.2). Thus, we have

$$\Phi(z_1) = P_{11}(z_1). \tag{1.3}$$

This is an interpolation constraint to be satisfied by all the admissible closed-loop transfer functions, irrespective of the selection of the controller. An analogous argument applies to the unstable zeros of $P_{21}(s)$.

The interpolation constraints of the type (1.3) are dictated by the requirement of internal stability. Construction of a norm-bounded analytic function $\Phi(s)$ satisfying the interpolation constraints (1.3) is exactly the subject of the classical interpolation theory [77][78]. The detailed exposition of the classical interpolation theory is found in the survey papers [69][16][55][56][105], and the monograph [4].

Let us consider next the MIMO case where both $P_{12}(s)$ and $P_{21}(s)$ are square and invertible. Let z_1 be an unstable zero of $P_{12}(s)$, that is, $\det P_{12}(z_1) = 0$, $\mathrm{Re}\ z_1 \geq 0$. In this case, we can find a vector $\xi_1 \neq 0$ such that

$$\xi_1^T P_{12}(z_1) = 0.$$

This relation implies

$$\xi_1^T \Phi(z_1) = \xi^T P_{11}(z_1), \tag{1.4}$$

which must be satisfied by every admissible closed-loop transfer function $\Phi(s)$. The identity (1.4) can be regarded as an interpolation constraint at $s = z_1$ in the direction of ξ_1^T. This type of interpolation problem was first investigated by Fedcina [27] and elaborated by the author in [52][53]. Although the relevance of interpolation theory to H^∞control was quite clear in terms of the plant transfer function $P(s)$ in the case that $P_{12}(s)$ and $P_{21}(s)$ are square, it was not an easy task to translate it into the language of state space. The notion of J-lossless conjugation introduced in [54] was a representation of the classical interpolation theory in the state space. The Pick matrix which plays a fundamental role in the classical interpolation theory is generalized and represented as a solution to the algebraic Riccati equation.

Actually, the J-lossless conjugation is more than a state space representation of the classical interpolation theory. It naturally leads to the notion of J-lossless factorization which represents a given matrix $G(s)$ as the product of a J-lossless matrix $\Theta(s)$ and a unimodular matrix $\Pi(s)$, that is,

$$G(s) = \Theta(s)\Pi(s). \tag{1.5}$$

The J-lossless factorization is an extension of the so-called J-spectral factorization which is written as

$$G^T(-s)JG(s) = \Pi^T(-s)J\Pi(s), \tag{1.6}$$

where J is a signature matrix. The relation (1.6) is easily derived from (1.5) using the relation $\Theta^T(-s)J\Theta(s) = J$ which is a defining property of the J-lossless matrix.

A salient feature of classical interpolation is its cascade structure. The Schur-Nevanlinna algorithm to solve the problem recursively results in the interpolants in the factorized form that was extensively used in circuit theory [38][104]. In the same vein, the J-lossless factorization (1.5) is closely related to the cascade structure of the closed-loop system which is exploited in the framework of the chain-scattering representation of the plant. The closed-loop transfer function (1.1) is written as

$$\Phi(s) = (G_{11}(s)K(s) + G_{12}(s))(G_{21}(s)K(s) + G_{22}(s))^{-1} \tag{1.7}$$

in terms of the chain-scattering representation $G(s)$ of the plant. It is easily seen that if $K(s)$ is represented in the same way as in (1.7), that is,

$$K(s) = (H_{11}(s)L(s) + H_{12}(s))(H_{21}(s)L(s) + H_{22}(s))^{-1}, \tag{1.8}$$

then substitution of (1.8) in (1.7) yields a similar type of expression for $\Phi(s)$ as

$$\Phi(s) = (J_{11}(s)L(s) + J_{12}(s))(J_{21}(s)L(s) + J_{22}(s))^{-1},$$

where $J(s)$ is the cascade connection of $G(s)$ and $H(s)$, that is,

$$J(s) = G(s)H(s).$$

This is what we call *cascade structure*. The chain-scattering representation of systems enables us to treat the feedback connection as a cascade connection. This framework makes everything much simpler and clarifies the fundamental structure in a straightforward way. This property of the chain-scattering representation has been extensively used in a variety of engineering fields to represent the scattering properties of physical systems. The factorization (1.5) is especially fit for exploiting the cascade structure. It should be noted that we made the assumption that both $P_{12}(s)$ and $P_{21}(s)$ were square and invertible to derive the interpolation condition (1.3). Extension to general cases was a serious issue in the mid 1980s.

In parallel with the approach by interpolation theory, there had been a method to solve the H^{∞}control problem based on the AAK theory [1]. This approach solved the problem under the same invertibility condition by reducing it to the Nehari extension problem [31]. For general cases, the problem is formulated as the *general distance problem* in this approach and was finally solved based on the unitary dilation which is a highly sophisticated mathematical machinery in operator theory [14].

It is a serious problem that the chain-scattering representation does not exist for nonsquare cases. This difficulty seems to be the main reason that the chain-scattering formalism has not been used as a theoretical framework for H^∞control, in spite of its suitability and compatibility for H^∞control of square plants. To circumvent this difficulty, a slightly different cascade structure was used in [35][95] and [96] which does not require the plant invertibility. This approach was successful in reducing the problem to the J-spectral factorization (1.6). The reasoning used in [35] and [96] look much more transparent than those in the approach of the general distance problem which uses a long string of problem reductions and a considerable amount of plant augmentations [31][34][3]. The J-spectral factorization approach in [35][96], however, is quite technical and the structure of H^∞control is still obscured. The meaning of factorization itself is not clear.

Our remedy taken in this book is to augment the plant in such a way that allows the chain-scattering representation and/or its dual form. More precisely, we append fictitious input u' and output y' to the original plant to make $P_{12}(s)$ or $P_{21}(s)$ square and invertible, as is shown in Figure 1.3. The chain-scattering representation can be obtained for those augmented plants and we can use the cascade structure for this representation. The crucial problem is that the synthesis of the controller should be done under the restriction that these fictitious inputs and/or outputs are not usable. We have found that a certain class of augmentations cannot improve the control performance from the viewpoint of H^∞control. In these augmentations, the fictitious input and output are useless for reducing the H^∞norm bound of the closed-loop transfer function. These augmentations are called *maximum* because the solutions of the Riccati equations associated with these augmentations are maximum in the solutions associated with other augmented plants. Identifying such a particular class of augmentations sheds new light on the fundamental structure of H^∞ control. This part is new and has not been published elsewhere.

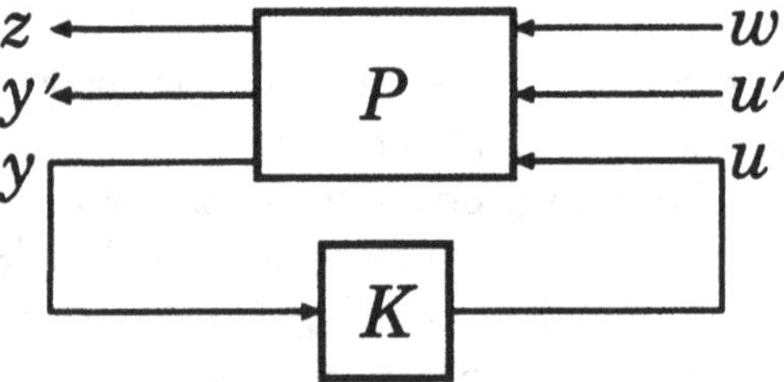

Figure 1.3 Augmented Plant.

Chapter 2

Elements of Linear System Theory

2.1 State-Space Description of Linear Systems

A linear system is described in the state space as

$$\begin{aligned} \dot{x} &= Ax + Bu, && (2.1a) \\ y &= Cx + Du, && (2.1b) \end{aligned}$$

where $u \in R^r$, $y \in R^m$, and $x \in R^n$ are the input, the output, and the state, respectively. The differential equation (2.1a) is usually referred to as a *state equation*, which is solved to be

$$x(t) = e^{At}x(0) + \int_0^t e^{A(t-\tau)}Bu(\tau)d\tau, \qquad (2.2)$$

where

$$e^{At} = I + \frac{At}{1!} + \frac{A^2t^2}{2!} + \cdots. \qquad (2.3)$$

Differentiation of (2.3) yields

$$\frac{d}{dt}e^{At} = Ae^{At}. \qquad (2.4)$$

In the description (2.1), the input $u(t)$ and the output $y(t)$ are *visible* externally, whereas the state $x(t)$ is not. The state is only accessible indirectly through the input $u(t)$ and the output $y(t)$. In this sense, the state has less physical reality than the input and the output. On the other hand, as was shown in (2.2), the state $x(0)$ at time $t = 0$ and the future input $\{u(t), t \geq 0\}$ determine the whole future behavior of the state $\{x(t), t \geq 0\}$. This implies that the state carries all the information about the past behavior of the system which is necessary to determine the future system behavior. If we know $x(0)$, we don't have to know anything else about the past behavior of the system, for example, $x(-1)$, to determine $x(t)$, $t \geq 0$. In this sense, the state is regarded as an *interface between the past and the future* with respect to the information carried over from the past to the future. From this point of view, we can say that the state is concerned with the *information* aspect rather than the *physical* aspect. Hence, we can choose the state in many different ways to represent physical aspects of the system, as long as it preserves the information. For instance, we can describe (2.1a) alternatively by a different frame of the state

$$x' = Tx, \tag{2.5}$$

where T is an arbitrary nonsingular matrix. The system (2.1) is then described by

$$\begin{aligned} \dot{x}' &= A'x' + B'u, & (2.6a)\\ y &= C'x + D'u, & (2.6b) \end{aligned}$$

where

$$A' = TAT^{-1}, \quad B' = TB, \quad C' = CT^{-1}, \quad D' = D. \tag{2.7}$$

The transformation (2.5) introduces a transformation (2.7) between the coefficient matrices of the state space descriptions. This transformation is called the *similarity transformation*, and the system (2.1) and the system (2.6) are said to be *similar* to each other. We sometimes write

$$\begin{bmatrix} A & B \\ C & D \end{bmatrix} \overset{T}{\sim} \begin{bmatrix} A' & B' \\ C' & D' \end{bmatrix},$$

if (2.7) holds.

Using an appropriate similarity transformation, we can represent the system (2.1) in various ways which might be convenient for some specific purposes. For instance, A' in (2.7) can be chosen as the Jordan canonical form:

$$A' = \begin{bmatrix} J_1 & 0 & \cdots & 0 \\ 0 & J_2 & \cdots & 0 \\ \cdots & & & \\ 0 & 0 & \cdots & J_k \end{bmatrix}, \quad J_i = \begin{bmatrix} \lambda_i & 1 & 0 & 0 \\ 0 & \lambda_i & 1 & 0 \\ \cdots & & & \\ 0 & 0 & 0 & \lambda_i \end{bmatrix},$$

or

$$A' = \begin{bmatrix} A_1 & 0 \\ 0 & A_2 \end{bmatrix},$$

where $A_i, i = 1, 2$, have specific spectrum configurations. One of the central themes of linear system theory is the canonical structure of linear systems, which amounts to finding the invariants of similarity transformations.

A linear system is described alternatively by the transfer function. Laplace transformation of both sides of (2.1) with $x(0) = 0$ yields

$$sX(s) = AX(s) + BU(s),$$
$$Y(s) = CX(s) + DU(s),$$

where the Laplace transform of each signal in (2.1) is denoted by the corresponding capital letter. Eliminating $X(s)$ in the preceding relations yields

$$Y(s) = G(s)U(s), \tag{2.8}$$
$$G(s) := C(sI - A)^{-1}B + D. \tag{2.9}$$

The matrix $G(s)$ relates $Y(s)$ with $U(s)$ linearly and is called a *transfer function* of the system (2.1). The relation (2.9) is usually represented as

$$G(s) := \left[\begin{array}{c|c} A & B \\ \hline C & D \end{array}\right]. \tag{2.10}$$

The preceding notation implies that $G(s)$ has a *realization* (2.1).

The transfer function represents the input/output relation generated by the system (2.1). The transfer function is an invariant of similarity transformation (2.7).

LEMMA 2.1 *The transfer function remains invariant with respect to the similarity transformation.*

Proof. Let $G'(s)$ be the transfer function of the system (2.6), Then, from (2.7), it follows that

$$\begin{aligned} G'(s) &= C'(sI - A')^{-1}B' + D' \\ &= CT^{-1}(sI - TAT^{-1})^{-1}TB + D \\ &= C(sI - A)^{-1}B + D = G(s) \end{aligned}$$

which establishes the assertion.

Now, a natural question arises whether the converse of the preceding lemma holds, that is, whether the two systems with the identical transfer function are similar to each other. This question is answered in the next section.

The invariance of the transfer functions with respect to similarity transformations is expressed as

$$\left[\begin{array}{c|c} A & B \\ \hline C & D \end{array}\right] = \left[\begin{array}{c|c} TAT^{-1} & TB \\ \hline CT^{-1} & D \end{array}\right]. \tag{2.11}$$

Before concluding the section, we collect state-space descriptions of various operations for transfer functions described in Figure 2.1.

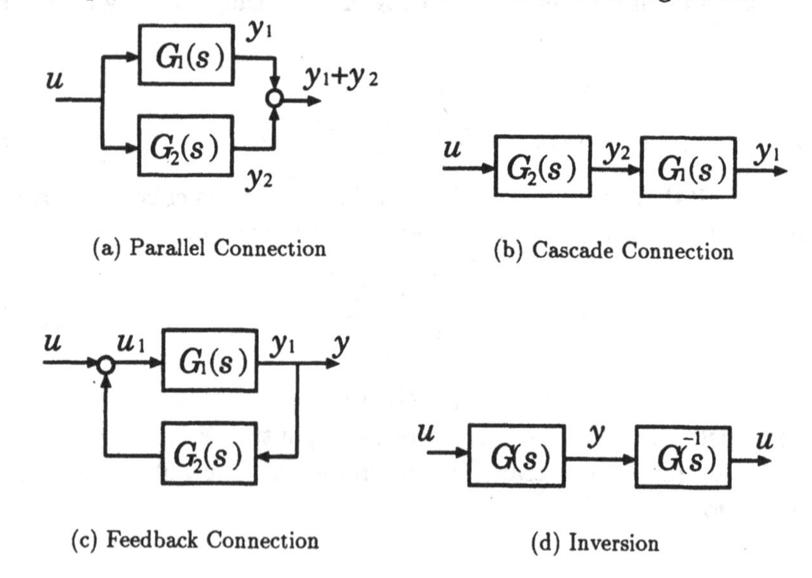

Figure 2.1 Connections of Transfer Functions.

We assume

$$G_i(s) = \left[\begin{array}{c|c} A_i & B_i \\ \hline C_i & D_i \end{array}\right], i = 1, 2.$$

(a) *Parallel Connection*

$$G(s) := G_1(s) + G_2(s) = \left[\begin{array}{cc|c} A_1 & 0 & B_1 \\ 0 & A_2 & B_2 \\ \hline C_1 & C_2 & D_1 + D_2 \end{array}\right]. \tag{2.12}$$

(b) *Cascade Connection*

$$G(s) := G_1(s)G_2(s) = \left[\begin{array}{cc|c} A_1 & B_1C_2 & B_1D_2 \\ 0 & A_2 & B_2 \\ \hline C_1 & D_1C_2 & D_1D_2 \end{array}\right]$$

$$= \left[\begin{array}{cc|c} A_2 & 0 & B_2 \\ B_1C_2 & A_1 & B_1D_2 \\ \hline D_1C_2 & C_1 & D_1D_2 \end{array}\right]. \tag{2.13}$$

(c) *Feedback Connection*

$$G(s) := (I - G_1(s)G_2(s))^{-1}G_1(s)$$

$$= \left[\begin{array}{cc|c} A_1 + B_1D_2VC_1 & B_1WC_2 & B_1W \\ B_2VC_1 & A_2 + B_2VD_1C_2 & B_2VD_1 \\ \hline VC_1 & VD_1C_2 & VD_1 \end{array}\right], \tag{2.14}$$

where $V := (I - D_1D_2)^{-1}$ and $W := (I - D_2D_1)^{-1}$.

(d) *Inversion*

$$G(s) := G(s)^{-1} = \left[\begin{array}{c|c} A - BD^{-1}C & BD^{-1} \\ \hline -D^{-1}C & D^{-1} \end{array}\right]. \tag{2.15}$$

(e) *Conjugate*

$$G^{\sim}(s) := G^T(-s) = \left[\begin{array}{c|c} -A^T & C^T \\ \hline -B^T & D^T \end{array}\right]. \tag{2.16}$$

2.2 Controllability and Observability

The notions of controllability and observability are of central importance in linear system theory and connect the state-space description (2.1) and transfer function (2.9). The duality between controllability and observability appears in various ways throughout linear system theory and manifests itself as one of the most salient characteristic features of linear systems.

Consider a linear system described by

$$\begin{bmatrix} \dot{x}_1 \\ \dot{x}_2 \end{bmatrix} = \begin{bmatrix} A_1 & A_{12} \\ 0 & A_2 \end{bmatrix} \begin{bmatrix} x_1 \\ x_2 \end{bmatrix} + \begin{bmatrix} B_1 \\ 0 \end{bmatrix} u, \tag{2.17a}$$

$$y = \begin{bmatrix} C_1 & C_2 \end{bmatrix} \begin{bmatrix} x_1 \\ x_2 \end{bmatrix}. \tag{2.17b}$$

Since $\dot{x}_2 = A_2 x_2$, this portion of the state is not affected by control input $u(t)$. In other words, the input $u(t)$ cannot control x_2 at all. Therefore, the system (2.17a) contains an uncontrollable part x_2 and such a system is called *uncontrollable.* A system that does not include an uncontrollable part is called *controllable.* More precisely, a system that cannot be expressed in the form (2.17a) by any similarity transformation is called controllable.

The preceding definition of controllability is indirect in the sense that it is defined by its negation. A more direct definition is as follows.

DEFINITION 2.2 *A system*

$$\dot{x} = Ax + Bu \tag{2.18}$$

is said to be controllable, *or the pair* (A, B) *is said to be* controllable, *if, for any initial condition* $x(0)$ *and* $t_1 > 0$, *there exists an input* $u(t)$ $(0 \leq t \leq t_1)$ *which brings the state at the origin, that is,* $x(t_1) = 0$.

The well-known characterizations of controllability are as follows.

LEMMA 2.3 *The following statements are equivalent.*

(i) (A, B) *is controllable.*

(ii) Rank $\begin{bmatrix} \lambda I - A & B \end{bmatrix} = n, \quad \forall \lambda.$

(iii) *(A, B) is not similar to any pair of the type*

$$\left(\begin{bmatrix} A_1 & A_{12} \\ 0 & A_2 \end{bmatrix}, \begin{bmatrix} B_1 \\ 0 \end{bmatrix}\right). \tag{2.19}$$

(iv) Rank $[B \quad AB \quad \cdots \quad A^{n-1}B] = n$.

(v) *If $\xi^T(sI - A)^{-1}B = 0$ for each s, then $\xi = 0$.*

(vi) *If $\xi^T e^{At}B = 0$ for $0 \le t < t_1$ with arbitrary $t_1 > 0$, then $\xi = 0$.*

Proof. (i) $\to$ (ii).
If (ii) does not hold, then rank $[\lambda I - A \quad B] < n$ for some λ. Therefore, there exists $\xi \neq 0$ such that

$$\xi^T[\lambda I - A \quad B] = 0.$$

From (2.18), it follows that

$$\xi^T \dot{x}(t) = \lambda \xi^T x(t),$$

for any input $u(t)$. Thus, $\xi^T x(t) = e^{\lambda t}\xi^T x(0)$. If $x(0)$ is chosen such that $\xi^T x(0) \neq 0$, then $\xi^T x(t) \neq 0$ for any $t \geq 0$. This implies that $x(t) \neq 0$ for any $u(t)$. Hence, (A, B) is not controllable.

(ii) $\to$ (iii).
Assume that there exists a similarity transformation T such that

$$TAT^{-1} = \begin{bmatrix} A_1 & A_{12} \\ 0 & A_2 \end{bmatrix}, \quad TB = \begin{bmatrix} B_1 \\ 0 \end{bmatrix}.$$

Let λ be an eigenvalue of A_2^T with eigenvector ξ_2. Clearly, we have

$$\begin{aligned} &[0 \quad \xi_2^T]T[\lambda I - A \quad B]\begin{bmatrix} T^{-1} & 0 \\ 0 & I \end{bmatrix} \\ &= [0 \quad \xi_2^T]\begin{bmatrix} \lambda I - A_1 & -A_{12} & B_1 \\ 0 & \lambda I - A_2 & 0 \end{bmatrix} = 0, \end{aligned}$$

which contradicts (ii).

(iii) $\rightarrow$ (iv).
Assume that rank $[B \quad AB \quad \cdots \quad A^{n-1}B] < n$. Then there exists a nonzero vector ξ such that $\xi^T B = \xi^T AB = \cdots = \xi^T A^{n-1}B = 0$. Let k be the maximum number such that $\{\xi, A^T\xi, \cdots, (A^T)^{k-1}\xi\}$ are linearly independent and let

$$T_2 = \begin{bmatrix} \xi^T \\ \xi^T A \\ \vdots \\ \xi^T A^{k-1} \end{bmatrix}.$$

Obviously, $k < n$, and $\xi^T A^k = \alpha_1 \xi^T + \alpha_2 \xi^T A + \cdots + \alpha_k \xi^T A^{k-1}$ for some real numbers $\alpha_1, \alpha_2, \cdots, \alpha_k$. Thus, we have

$$T_2 A = A_2 T_2,$$
$$A_2 = \begin{bmatrix} 0 & 1 & 0 & \cdots & 0 \\ 0 & 0 & 1 & \cdots & 0 \\ \vdots & \vdots & \vdots & \ddots & \vdots \\ \alpha_1 & \alpha_2 & \alpha_3 & \cdots & \alpha_k \end{bmatrix}.$$

Let T_1 be any matrix in $R^{(n-k)\times n}$ such that

$$T := \begin{bmatrix} T_1 \\ T_2 \end{bmatrix}$$

is nonsingular. Since $T_2 B = 0$, we can easily see that

$$TA = \begin{bmatrix} A_1 & A_{12} \\ 0 & A_2 \end{bmatrix} T, \quad TB = \begin{bmatrix} B_1 \\ 0 \end{bmatrix},$$

for some A_1, A_{12} and B_1. This contradicts (iii).

(iv) $\rightarrow$ (v).
If $\xi^T(sI - A)^{-1}B = 0 \quad \forall s$ for some $\xi \neq 0$, the Laurent expansion of $(sI - A)^{-1}$ implies

$$(\xi^T s^{-1} + \xi^T A s^{-2} + \cdots + \xi^T A^k s^{-(k+1)} + \cdots)B = 0, \quad \forall s.$$

This obviously implies that

$$\xi^T A^k B = 0, \quad \forall k,$$

which contradicts (iv).

(v) → (vi).
Assume that $\xi^T e^{At} B \equiv 0$ for $0 \le t < t_1$. Then, $d^k(e^{At}B)/dt^k = 0$ at $t = 0$ implies

$$\xi^T A^k B = 0, \quad \forall k.$$

This obviously implies $\xi^T(sI - A)^{-1}B = 0$ for each s.

(vi) → (i).
Let

$$M(t_1) := \int_0^{t_1} e^{-At} BB^T e^{-A^T t} dt.$$

If (vi) holds, $M(t_1)$ is nonsingular for any $t_1 > 0$. Indeed, if $M(t_1)\xi = 0$, we have

$$\xi^T M(t_1)\xi = \int_0^{t_1} \|\xi^T e^{-At} B\|^2 dt = 0.$$

This implies that $\xi^T e^{-At} B = 0$ for $0 \le t < t_1$. Hence, $\xi = 0$.
Let

$$u(t) := -B^T e^{-A^T t} M(t_1)^{-1} x(0). \tag{2.20}$$

This input brings the state of (2.1a) to the origin at $t = t_1$. Indeed, we see that

$$\begin{aligned} x(t_1) &= e^{At_1} x(0) + \int_0^{t_1} e^{A(t_1 - t)} Bu(t) dt \\ &= e^{At_1} \left(x(0) - \int_0^{t_1} e^{-At} BB^T e^{-A^T t} dt \cdot M(t_1)^{-1} x(0) \right) \\ &= 0, \end{aligned}$$

which establishes the assertion. ∎

If (A, B) is controllable, we can bring the state of the system (2.1a) to any position in the state-space at an arbitrary time starting at the origin. The term controllability was originated from this fact.

The dual notion of controllability is observability, which is defined as follows.

DEFINITION 2.4 *A system*

$$\dot{x} = Ax, \quad y = Cx \tag{2.21}$$

is said to be observable, or the pair (A, C) is said to be observable, if the output segment $\{y(t); 0 \leq t < t_1\}$ of arbitrary length uniquely determines the initial state $x(0)$ of (2.21).

In (2.21), the input $u(t)$ is not included. The preceding definition can be extended in an obvious way to the case where input is incorporated.

A typical example of an unobservable system is described as

$$\begin{bmatrix} \dot{x}_1 \\ \dot{x}_2 \end{bmatrix} = \begin{bmatrix} A_2 & 0 \\ A_{21} & A_2 \end{bmatrix} \begin{bmatrix} x_1 \\ x_2 \end{bmatrix} + \begin{bmatrix} B_1 \\ B_2 \end{bmatrix} u, \tag{2.22a}$$

$$y = \begin{bmatrix} C_1 & 0 \end{bmatrix} \begin{bmatrix} x_1 \\ x_2 \end{bmatrix}. \tag{2.22b}$$

It is clear that the component x_2 of the state cannot affect $y(t)$ and hence cannot be estimated from the observation of $y(t)$.

The duality between the representations (2.17) and (2.22) is obvious. Actually, it is remarkable that the observability is characterized in the way completely dual to the controllability, as is shown in the following results:

LEMMA 2.5 *The following statements are all equivalent:*

(i) *(A, C) is controllable;*

(ii) rank $\begin{bmatrix} \lambda I - A \\ C \end{bmatrix} = n, \quad \forall \lambda;$

(iii) *(A, C) is not similar to any pair of the type*

$$\left(\begin{bmatrix} A_1 & 0 \\ A_{21} & A_2 \end{bmatrix}, \; [C_1 \;\; 0] \right);$$

(iv) rank $\begin{bmatrix} C \\ CA \\ \vdots \\ CA^{n-1} \end{bmatrix} = n;$

(v) *if $C(sI - A)^{-1}\eta = 0$ for each s, then $\eta = 0$;*

(vi) *if* $Ce^{At}\eta = 0$ *for* $0 \leq t < t_1$ *with arbitrary* $t_1 > 0$, *then* $\eta = 0$.

The proof can be done in a similar way to that of Lemma 2.3 and is left to the reader.

If (A_1, B_1) is controllable in (2.17), then x_1 represents the controllable portion of the state, whereas x_2 denotes its uncontrollable portion. In this sense, the system (2.1a) is decomposed into the controllable portion and the uncontrollable one in (2.17a). The representation (2.17a) is schematically represented in Figure 2.2.

In the dual way, the representation (2.22) decomposes the state into the observable and the unobservable portions, provided that the pair (A_1, C_1) is observable. The schematic representation of the system (2.22) is given in Figure 2.3.

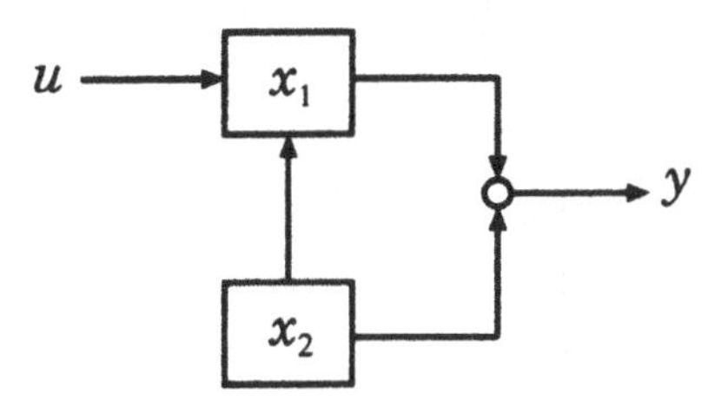

Figure 2.2 Controllable and Uncontrollable Portions.

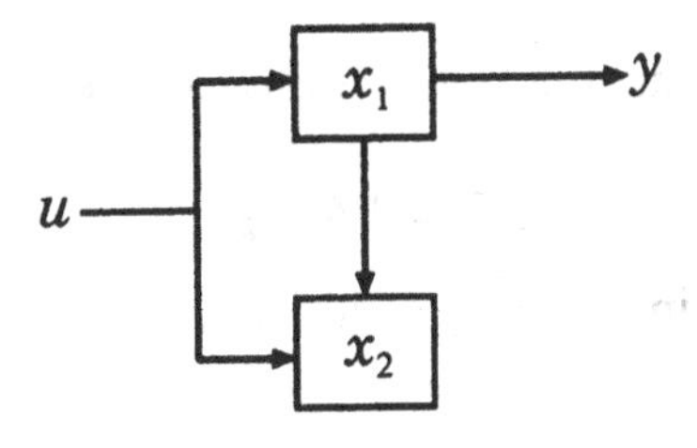

Figure 2.3 Observable and Unobservable Portions.

If (A, B) is controllable and (A, C) is observable, then the system (2.1) is said to be *minimal* or *irreducible*. The converse of Lemma 2.1 holds under the assumption that the systems under consideration are minimal. The following lemma is one of the most fundamental results in linear system theory.

LEMMA 2.6 *If two minimal realizations have an identical transfer* *function, then they are similar to each other. In other words, if*

$$\left[\begin{array}{c|c} A_1 & B_1 \\ \hline C_1 & D_1 \end{array}\right] = \left[\begin{array}{c|c} A_2 & B_2 \\ \hline C_2 & D_2 \end{array}\right] \tag{2.23}$$

and both realizations are minimal, then

$$\begin{bmatrix} A_1 & B_1 \\ C_1 & D_1 \end{bmatrix} \overset{T}{\sim} \begin{bmatrix} A_2 & B_2 \\ C_2 & D_2 \end{bmatrix}. \tag{2.24}$$

for some nonsingular T.

Proof. Let n be the size of A_1 and A_2 and define

$$M_i = [B_i \quad A_iB_i \quad A_i^2B_i \quad \cdots \quad A_i^{n-1}B_i], \quad i = 1, 2$$

$$W_i = \begin{bmatrix} C_i \\ C_iA_i \\ \vdots \\ C_iA_i^{n-1} \end{bmatrix}, \quad i = 1, 2.$$

Since $D_1 + C_1(sI - A_1)^{-1}B_1 = D_2 + C_2(sI - A_2)^{-1}B_2$ for each s, we have

$$D_1 = D_2, \quad C_1A_1^kB_1 = C_2A_2^kB_2, \quad \forall k,$$

by comparing each coefficient of the Laurent expansion of $(sI - A_i)^{-1}$, $i = 1, 2$. In other words,

$$W_1M_1 = W_2M_2. \tag{2.25}$$

Since both realizations in (2.23) are minimal, both $M_1M_1^T$ and $W_1^TW_1$ are invertible. Let

$$T_1 := M_2M_1^T(M_1M_1^T)^{-1}, \quad T_2 := (W_1^TW_1)^{-1}W_1^TW_2.$$

Due to (2.25), $T_2T_1 = (W_1^TW_1)^{-1}W_1^TW_2M_2M_1^T(M_1M_1^T)^{-1} = I$. Therefore, $T_2 = T_1^{-1}$. Also,

$$T_2M_2 = M_1, \quad W_2T_1 = W_1,$$

which implies that

$$B_1 = T_2B_2, \quad C_1 = C_2T_2^{-1}.$$

Since $W_1A_1M_1 = W_2A_2M_2 = W_1T_1^{-1}A_2T_1M_1$, we have

$$A_1 = T_1^{-1}A_2T_1,$$

which establishes (2.24) with $T = T_1$.

The description (2.17a) decomposes the state-space into the controllable portion and the uncontrollable one, whereas the description (2.22a) decomposes the state-space into the observable portion and the unobservable one. We can integrate these two decompositions to get the *canonical decomposition*, based on the controllability matrix M and the observability matrix W given, respectively, by

$$M := [B \quad AB \quad \cdots \quad A^{n-1}B],$$

$$W := \begin{bmatrix} C \\ CA \\ \vdots \\ CA^{n-1} \end{bmatrix},$$

which were used in the proof of Lemma 2.6.

We introduce the following four subspaces in $\mathbf{R}^n$:

$$R_a := M\ \mathrm{Ker}WM = \mathrm{Im}M \cap \mathrm{Ker}W, \tag{2.26}$$

$$R_b := \mathrm{Im}M \cap (\mathrm{Im}W^T + \mathrm{Ker}M^T), \tag{2.27}$$

$$R_c := \mathrm{Ker}W \cap (\mathrm{Im}W^T + \mathrm{Ker}M^T), \tag{2.28}$$

$$R_d := W^T\ \mathrm{Ker}M^TW^T = \mathrm{Im}W^T \cap \mathrm{Ker}M^T. \tag{2.29}$$

We see that

$$\mathrm{Im}M = R_a + R_b, \tag{2.30}$$

$$\mathrm{Ker}W = R_a + R_c. \tag{2.31}$$

Indeed, since $\mathbf{R}^n = \mathrm{Im}W^T + \mathrm{Ker}M^T + \mathrm{Ker}W \cap \mathrm{Im}M$, $\mathrm{Im}M = \mathrm{Im}M \cap (\mathrm{Im}W^T + \mathrm{Ker}M^T + \mathrm{Ker}W \cap \mathrm{Im}M) = \mathrm{Im}M \cap (\mathrm{Im}W^T + \mathrm{Ker}M^T) + \mathrm{Im}M \cap (\mathrm{Ker}W \cap \mathrm{Im}M) = R_b + R_a$. The relation (2.31) can be proven similarly. We also see that $R_a + R_b + R_c = \mathrm{Im}M + \mathrm{Ker}W$ and R_d is the orthogonal complement of $\mathrm{Im}M + \mathrm{Ker}W$. Hence,

$$\mathbf{R}^n = R_a + R_b + R_c + R_d. \tag{2.32}$$

Obviously, $\mathrm{Im}M$ and $\mathrm{Ker}W$ are A-invariant. Hence,

$$AR_a \subset R_a, \tag{2.33}$$

$$AR_b \subset \mathrm{Im}M = R_a + R_b, \tag{2.34}$$

$$AR_c \subset \mathrm{Ker}W = R_a + R_c. \tag{2.35}$$

Also, we have

$$\mathrm{Im}B \subset \mathrm{Im}M = R_a + R_b, \tag{2.36}$$

$$\mathrm{Ker}C \supset \mathrm{Ker}W = R_a + R_c\ . \tag{2.37}$$

Let T_a, T_b, T_c, and T_d be matrices such that their columns span R_a, R_b, R_c, and R_d, respectively. The relations (2.33)~(2.37) imply the following representations:

$$A\begin{bmatrix} T_a & T_b & T_c & T_d \end{bmatrix} = \begin{bmatrix} T_a & T_b & T_c & T_d \end{bmatrix}\begin{bmatrix} A_{aa} & A_{ab} & A_{ac} & A_{ad} \\ 0 & A_{bb} & 0 & A_{bd} \\ 0 & 0 & A_{cc} & A_{cd} \\ 0 & 0 & 0 & A_{dd} \end{bmatrix},$$

$$B = \begin{bmatrix} T_a & T_b & T_c & T_d \end{bmatrix}\begin{bmatrix} B_a \\ B_b \\ 0 \\ 0 \end{bmatrix},$$

$$C\begin{bmatrix} T_a & T_b & T_c & T_d \end{bmatrix} = \begin{bmatrix} 0 & C_b & 0 & C_d \end{bmatrix}.$$

Due to (2.32), $T := \begin{bmatrix} T_1 & T_2 & T_3 & T_4 \end{bmatrix} \in \mathbf{R}^{n\times n}$ is invertible. Therefore, we have the following fundamental result.

THEOREM 2.7 [47] *Every state-space representation is similar to the following structure*

$$\frac{d}{dt}\begin{bmatrix} x_a \\ x_b \\ x_c \\ x_d \end{bmatrix} = \begin{bmatrix} A_{aa} & A_{ab} & A_{ac} & A_{ad} \\ 0 & A_{bb} & 0 & A_{bd} \\ 0 & 0 & A_{cc} & A_{cd} \\ 0 & 0 & 0 & A_{dd} \end{bmatrix}\begin{bmatrix} x_a \\ x_b \\ x_c \\ x_d \end{bmatrix} + \begin{bmatrix} B_a \\ B_b \\ 0 \\ 0 \end{bmatrix} u \tag{2.38a}$$

$$y = [0 \quad C_b \quad 0 \quad C_d] \begin{bmatrix} x_a \\ x_b \\ x_c \\ x_d \end{bmatrix} + Du. \tag{2.38b}$$

Each component of the state has the following meaning,

x_a : *controllable but unobservable portion,*
x_b : *controllable and observable portion,*
x_c : *uncontrollable and unobservable portion,*
x_d : *uncontrollable but observable portion.*

The representation (2.38) is often referred to as a *canonical decomposition*. Its schematic diagram is shown in Figure 2.4.

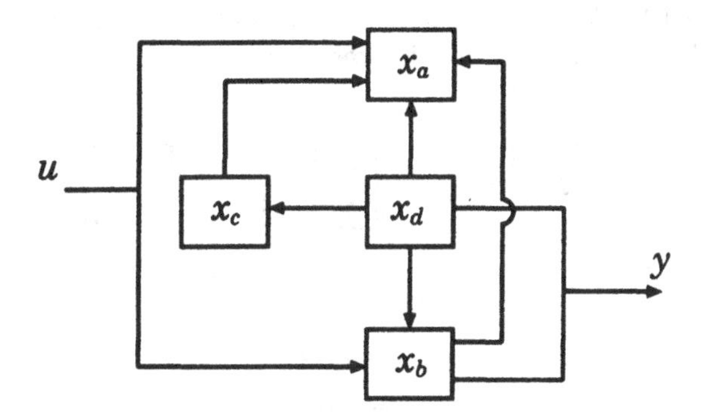

Figure 2.4 Canonical Decomposition.

In the preceding diagram, the only pass from the input u to the output y is through x_b. This implies that the transfer function $G(s)$ which represents the input/output relation of the system involves only the x_b-portion, that is, the controllable and observable portion. This is confirmed by the identity

$$\left[\begin{array}{cccc|c} A_{aa} & A_{ab} & A_{ac} & A_{ad} & B_a \\ 0 & A_{bb} & 0 & A_{bd} & B_b \\ 0 & 0 & A_{cc} & A_{cd} & 0 \\ 0 & 0 & 0 & A_{dd} & 0 \\ \hline 0 & C_b & 0 & C_d & D \end{array}\right] = \left[\begin{array}{c|c} A_{bb} & B_b \\ \hline C_b & D \end{array}\right]$$

$$= D + C_b(sI - A_{bb})^{-1}B_b.$$

2.3 State Feedback and Output Insertion

If the input u in (2.1a) is determined as an affine function of the state, that is,

$$u = Fx + u', \tag{2.39}$$

then u is called the *state feedback*. The additional control input u' can be determined for other purposes. Substitution of (2.39) in (2.1a) yields

$$\dot{x} = (A + BF)x + Bu'. \tag{2.40}$$

Thus, the state feedback (2.39) changes the A-matrix of the system from A to $A + BF$. Figure 2.5 shows a block diagram of the state feedback (2.39). As was stated in Section 2.1, the state vector contains enough information about the past to determine the future behavior of the system. In this sense, the state feedback represents an ideal situation because it can utilize the full information for feedback.

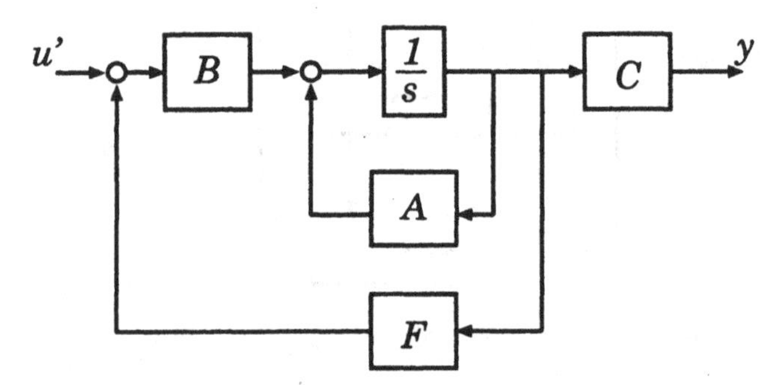

Figure 2.5 Block-Diagram of State Feedback.

The eigenvalues of A (denoted by $\sigma(A)$) are important factors that determine the behavior of the system (2.1). They are the singular points of the transfer function $G(s)$ and are called the *poles* of $G(s)$. The eigenvalues of $A+BF$ are the so-called *closed-loop poles* with respect to a state feedback (2.39). To choose the configuration of the closed-loop poles is one of the primal objectives of state feedback.

DEFINITION 2.8 *The pair (A, B) is said to be pole-assignable, if, for any set of complex numbers $\Lambda = \{\lambda_1, \lambda_2, \cdots, \lambda_n\}$, there exists a state feedback (2.39) such that $\sigma(A + BF) = \Lambda$.*

The following fact holds which is one of the most fundamental results in linear system theory.

THEOREM 2.9 *The following statements are equivalent.*

(i) *(A, B) is pole-assignable.*

(ii) *(A, B) is controllable*

Proof. (i) $\rightarrow$ (ii).
If (A, B) is not controllable, rank $\begin{bmatrix} \lambda I - A & B \end{bmatrix} < n$ for some λ due to Lemma 2.3. Hence, $\xi^T(\lambda I - A) = 0$, $\xi^T B = 0$ for some $\xi \neq 0$. Therefore, for any F, $\xi^T(\lambda I - A - BF) = 0$. This implies that λ is an eigenvalue of $A + BF$ for each F. Hence, (A, B) is not pole-assignable.

(ii)$\rightarrow$(i).
We first prove the assertion for the case with scalar input; that is, $B = b \in \mathbf{R}^{n \times 1}$. In that case, (ii) implies that $\{b, Ab, \cdots, A^{n-1}b\}$ is linearly independent. Let $\{\lambda_1, \lambda_2, \cdots, \lambda_n\}$ be an arbitrary set of n distinct complex numbers which are not the eigenvalues of A. We show that

$$f_i := (\lambda_i I - A)^{-1} b, \quad i = 1, 2, \cdots, n \tag{2.41}$$

is linearly independent. Indeed, we see that

$$(\lambda I - A)^{-1} b = (\psi_n(\lambda) A^{n-1} + \psi_{n-1}(\lambda) A^{n-2} + \cdots + \psi_1(\lambda) I) b / \psi_0(\lambda), \tag{2.42}$$

where $\psi_i(\lambda), i = 0, \cdots, n$ is the sequence of polynomials of degree i defined sequentially as follows:

$$\begin{aligned} \psi_0(\lambda) &= \det(\lambda I - A), \\ \psi_i(\lambda) &= (\psi_{i-1}(\lambda) - \psi_{i-1}(0))/\lambda, \quad i = 1, \cdots, n, \\ \psi_n(\lambda) &= 1. \end{aligned} \tag{2.43}$$

This implies

$$\begin{bmatrix} f_1 & f_2 & \cdots & f_n \end{bmatrix} = \begin{bmatrix} A^{n-1}b & A^{n-2}b & \cdots & Ab & b \end{bmatrix} \Theta \Psi,$$

$$\Theta := \begin{bmatrix} \psi_n(\lambda_1) & \psi_n(\lambda_2) & \cdots & \psi_n(\lambda_n) \\ \psi_{n-1}(\lambda_1) & \psi_{n-1}(\lambda_2) & \cdots & \psi_{n-1}(\lambda_n) \\ \vdots & \vdots & & \vdots \\ \psi_1(\lambda_1) & \psi_1(\lambda_2) & \cdots & \psi_1(\lambda_n) \end{bmatrix},$$

$$\Psi := \begin{bmatrix} 1/\psi_0(\lambda_1) & 0 & \cdots & 0 \\ 0 & 1/\psi_0(\lambda_2) & \cdots & 0 \\ \vdots & \vdots & \ddots & \vdots \\ 0 & 0 & \cdots & 1/\psi_0(\lambda_n) \end{bmatrix}.$$

Actually, the matrix Θ is calculated to be

$$\Theta = \begin{bmatrix} 1 & 0 & \cdots & 0 \\ \beta_1 & 1 & \cdots & 0 \\ \vdots & \vdots & \ddots & \vdots \\ \beta_{n-1} & \beta_{n-2} & \cdots & 1 \end{bmatrix} \begin{bmatrix} 1 & 1 & \cdots & 1 \\ \lambda_1 & \lambda_2 & \cdots & \lambda_n \\ \vdots & \vdots & \ddots & \vdots \\ \lambda_1^{n-1} & \lambda_2^{n-1} & \cdots & \lambda_n^{n-1} \end{bmatrix}, \tag{2.44}$$

where $\psi_0(\lambda) = \det(\lambda I - A) = \lambda^n + \beta_1\lambda^{n-1} + \cdots + \beta_n$. Thus, Θ is nonsingular for distinct λ_i. Therefore, we conclude that f_i is independent. Let F be a $1 \times n$ matrix such that

$$Ff_i = 1, \quad i = 1, 2, \cdots, n$$

or, equivalently,

$$F = \begin{bmatrix} 1 & 1 & \cdots & 1 \end{bmatrix} \begin{bmatrix} f_1 & f_2 & \cdots & f_n \end{bmatrix}^{-1}.$$

Due to (2.41),

$$(\lambda_i I - A - bF)f_i = b - b = 0.$$

This implies that λ_i is an eigenvalue of $A + bF$. Since λ_i is arbitrary, the assertion has been proven for the scalar input case.

The multi-input case can be reduced to the scalar-input case by utilizing the fact that if (A, B) is controllable, we can find $\overline{F}$ and $\overline{g}$ such that $(A + B\overline{F}, B\overline{g})$ is controllable. ∎

In the case where the system is not controllable, it is desirable that the uncontrollable mode is stable, that is, the matrix A_2 in (2.17a) is stable. In that case, the system is made stable by application of suitable state feedback.

DEFINITION 2.10 *The pair (A, B) is said to be stabilizable, if there exists a matrix F such that $A+BF$ is stable.*

Obviously, a controllable pair is stabilizable, but the converse is not true.

The dual of the state feedback (2.39) is the *output insertion* which is represented as

$$\dot{x} = Ax + Bu + Ly + \xi', \tag{2.45}$$

where ξ' is an auxiliary signal to be used for other purposes. A well-known identity state observer corresponds to the case where $\xi' = LCx$. The "closed-loop" poles are given by the eigenvalues of $A + LC$. Figure 2.6 shows a block diagram of output insertion (2.45). The dualization of Theorem 2.9 shows that the eigenvalues of $A + LC$ can be chosen arbitrarily by choosing L, if and only if (A, C) is observable.

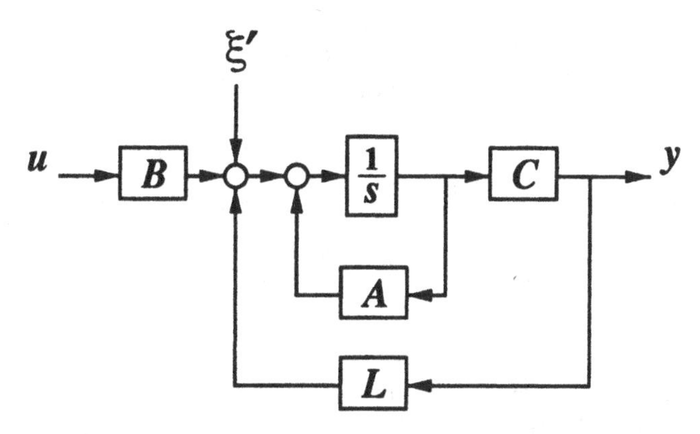

Figure 2.6 Block-Diagram of Output Insertion.

Through the output insertion, we can define the dual of stabilizable pair.

DEFINITION 2.11 *The pair (A, C) is said to be detectable if there exists a matrix L such that $A+LC$ is stable.*

Obviously, an observable pair is detectable, but the converse is not true.

The values of s at which $G(s)$ loses its normal rank are called the *transmission zeros* or the *zeros* of the system (2.1). If the realization (2.1) is minimal, z is a transmission zero of $G(s)$ iff

$$\operatorname{rank} \begin{bmatrix} A - zI & B \\ C & D \end{bmatrix} < n + \min(m, r). \tag{2.46}$$

The transmission zeros are important factors for control system design. Sometimes, they are more important than the poles. For instance, the existence of right-half plane zeros severely limits the achievable closed-loop performance. It is important to notice that the transmission zeros are invariant with respect to the state feedback and the input transformation

$$u = Fx + Uu', \quad U : \text{invertible}. \tag{2.47}$$

Indeed, application of (2.47) to (2.1) yields

$$\dot{x} = (A + BF)x + BUu', \tag{2.48a}$$

$$y = (C + DF)x + DUu'. \tag{2.48b}$$

It follows that

$$\begin{aligned}
&\operatorname{rank}\begin{bmatrix} A + BF - \lambda I & BU \\ C + DF & DU \end{bmatrix} \\
&= \operatorname{rank}\left(\begin{bmatrix} A - \lambda I & B \\ C & D \end{bmatrix}\begin{bmatrix} I & 0 \\ F & U \end{bmatrix}\right) \\
&= \operatorname{rank}\begin{bmatrix} A - \lambda I & B \\ C & D \end{bmatrix},
\end{aligned}$$

which shows the invariance of zeros.

Also, the transmission zeros are invariant under the output insertion and output transformation

$$\begin{aligned}
\dot{x} &= Ax + Bu + Ly \\
y' &= Vy, \qquad V : \text{invertible},
\end{aligned}$$

which gives rise to the state-space representation

$$\dot{x} = (A + LC)x + (B + LD)u, \tag{2.49a}$$

$$y' = V(Cx + Du). \tag{2.49b}$$

It follows that

$$\begin{aligned}
&\operatorname{rank}\begin{bmatrix} A + LC - \lambda I & B + LD \\ VC & VD \end{bmatrix} \\
&= \operatorname{rank}\begin{bmatrix} I & L \\ 0 & V \end{bmatrix}\begin{bmatrix} A - \lambda I & B \\ C & D \end{bmatrix} \\
&= \operatorname{rank}\begin{bmatrix} A - \lambda I & B \\ C & D \end{bmatrix},
\end{aligned}$$

which shows the invariance of zeros.

We summarize the preceding reasoning as follows,

LEMMA 2.12 *The transmission zeros of the transfer function are invariant with respect to state feedback, input transformation, output insertion, and output transformation.*

2.4 Stability of Linear Systems

Consider an autonomous linear system obtained by putting $u \equiv 0$ in (2.1a),

$$\dot{x} = Ax. \tag{2.50}$$

This system is said to be *asymptotically stable* if the solution $x(t)$ converges to the origin as $t \to \infty$ for each initial state $x(0)$. Since the solution of (2.50) is given by $x(t) = e^{At}x(0)$, the system (2.50) is asymptotically stable iff

$$\|e^{At}\| \to 0, \quad \text{as} \quad t \to \infty. \tag{2.51}$$

If (2.51) holds, A is said to be *stable* or *Hurwitz*. It is easily shown that A is stable iff each eigenvalue of A has a negative real part, that is,

$$\text{Re}\,[\lambda_i(A)] < 0, \quad i = 1, \cdots, n. \tag{2.52}$$

A simple method of checking the stability of A is known as the Lyapunov theorem which reduces the stability problem to solving a linear algebraic equation with respect to P given by

$$PA + A^T P = -Q. \tag{2.53}$$

The preceding equation is called a *Lyapunov equation*.

LEMMA 2.13 *The following statements are equivalent.*

(i) *A is stable.*

(ii) *There exists a positive number σ and M such that*

$$\|e^{At}\| \leq Me^{-\sigma t} \tag{2.54}$$

(iii) *For each matrix $Q > 0$, the Lyapunov equation (2.53) has a solution $P > 0$.*

Proof. (i) $\rightarrow$ (ii).
Since A is stable, there exists $\sigma > 0$ such that $-\sigma > \mathrm{Re}[\lambda_i(A)]$ for $i = 1, 2, \cdots, n$. In other words, $\sigma I + A$ is stable. Since $\|e^{(\sigma I + A)t}\| \rightarrow 0$, $\|e^{(\sigma I+A)t}\|$ is bounded for each t. Therefore, the inequality

$$\|e^{(\sigma I+A)t}\| \leq M \tag{2.55}$$

holds for some $M > 0$, which implies (2.54).

(ii) $\rightarrow$ (iii).
Due to (2.54), we can define

$$P := \int_0^\infty e^{A^T t} Q e^{At} dt.$$

Since $Q > 0$, $P > 0$. It is easy to see that P is a solution to Equation (2.53). Indeed,

$$\begin{aligned} PA + A^T P &= \int_0^\infty e^{A^T t} Q e^{At} A dt + \int_0^\infty A^T e^{A^T t} Q e^{At} dt \\ &= \int_0^\infty \frac{d}{dt}(e^{A^T t} Q e^{At}) dt = [e^{A^T t} Q e^{At}]_0^\infty \\ &= -Q. \end{aligned}$$

(iii) $\rightarrow$ (i).
Let λ be an eigenvalue of A and ξ the corresponding eigenvector, that is, $A\xi = \lambda\xi$. The premultiplication by ξ^* and the postmultiplication by ξ of (2.53) yield

$$(\lambda + \overline{\lambda})\xi^* P \xi = -\xi^* Q \xi.$$

Since both P and Q are positive definite from the assumption, we conclude that $\lambda + \overline{\lambda} < 0$. Hence, $\mathrm{Re}[\lambda] < 0$, which establishes the assertion.

∎

We sometimes use a positive semidefinite Q in (2.53), instead of a positive definite one. A typical example is the case $Q = C^T C$.

LEMMA 2.14 *Assume that (A, C) is observable. The Lyapunov equation*

$$PA + A^T P = -C^T C \tag{2.56}$$

has a positive definite solution iff A is stable.

Proof. If A is stable, the solution to (2.56) is given by

$$P = \int_0^\infty e^{A^T t} C^T C e^{At} dt \geq 0.$$

If $Px = 0$ for some x, we see that

$$\int_0^\infty \|Ce^{At}x\|^2 dt = 0.$$

Hence, $Ce^{At}x = 0, \forall t$. Due to Lemma 2.5(vi), $x = 0$. Hence, $P > 0$. Conversely, assume that the solution P of (2.56) is positive definite. Let λ be an eigenvalue of A with corresponding eigenvector ξ, that is, $A\xi = \lambda\xi$. As in the proof of Lemma 2.13, we have

$$(\lambda + \overline{\lambda}) x^* P x = -\|Cx\|^2.$$

Since (A, C) is observable, $Cx \neq 0$. Hence, $\lambda + \overline{\lambda} < 0$, which verifies the assertion. ∎

We can state the converse of the preceding lemma.

LEMMA 2.15 *If the Lyapunov equation (2.56) has a positive definite solution for a stable A, then (A, C) is observable.*

Proof. Assume that (A, C) is not observable. Then there exist λ and x such that $Ax = \lambda x$, $Cx = 0$. From (2.56), it follows that $\bar{x}^T P x \cdot (\lambda + \bar{\lambda}) = 0$. If A is stable, $\lambda + \bar{\lambda} < 0$. Since $P > 0$, we conclude that $x = 0$. ∎

A stronger version of Lemma 2.14 is given as follows.

LEMMA 2.16 *Assume that (A, C) is detectable. The Lyapunov equation (2.56) has a positive semi-definite solution iff A is stable.*

The proof is similar to that of Lemma 2.14 and is omitted here.

The solution P of (2.56) is called the *observability Gramian* of the system (2.1). Its dual is the *controllability Gramian* defined as the solution of

$$PA^T + AP = -BB^T. \tag{2.57}$$

Assuming that A is stable, the controllability Gramian is positive definite, iff (A, B) is controllable.

Problems

[1] Assume that the eigenvalues of A are all distinct. Prove that if (A, B) is controllable, there exists a vector $g \in \mathbf{R}^r$ such that (A, Bg) is controllable.

[2] Show that if (A, B) is controllable, then there exist a matrix F and a vector g such that $(A+BF, Bg)$ is controllable.

[3] In the proof of Lemma 2.6, the sizes of A_1 and A_2 are assumed to be the same from the outset. Justify this assumption.

[4] Show that the sequence $\psi_i(\lambda)$ defined in (2.43) satisfies the identity

$$\psi_0(\lambda) = \psi_n(0)\lambda^n + \psi_{n-1}(0)\lambda^{n-1} + \cdots + \psi_0(0).$$

[5] Using the identity $\psi_0(A) = 0$ (Caley-Hamilton theorem), prove the identity (2.42).

[6] Show that the following two statements are identical.

(i) (A, B) is stabilizable.

(ii) $\text{rank} \begin{bmatrix} \lambda I - A & B \end{bmatrix} = n, \quad \forall \text{Re}\, \lambda \geq 0.$

[7] Write down a state-space model of the electrical circuit in Figure 2.7 taking the terminal current as the input and the voltage as the output. Show that the state space form is uncontrollable if

$$R_1 R_2 = \frac{L}{C}.$$

In this case, derive the canonical decomposition of the state-space form. Compute the transfer function $Z(s)$ of the state-space form and show that $Z(s) = R_1$ in the case

$$R_1 = R_2 = \sqrt{\frac{L}{C}}.$$

Give a dynamical interpretation of this condition.

[8] In the proof of (2.30), the following facts were used. Show them.

(i) For any matrix U with n rows,

$$\mathbf{R}^n = \mathrm{Im} U + \mathrm{Ker} U^T.$$

(ii) For any subspaces R_1, R_2, and R_3 with $R_1 \supset R_3$,

$$R_1 \cap (R_2 + R_3) = R_1 \cap R_2 + R_3.$$

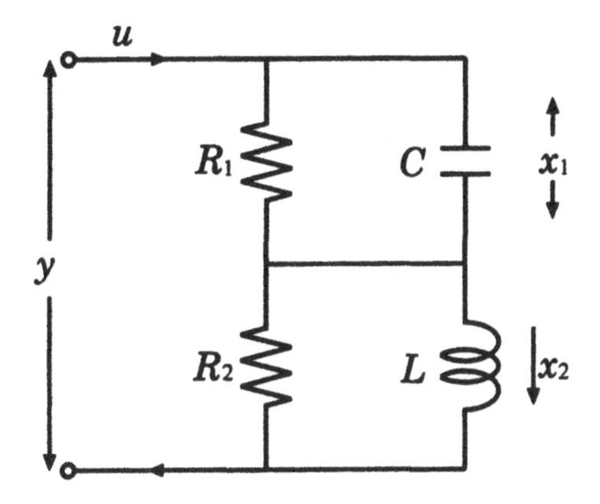

Figure 2.7 A one-terminal electrical circuit.

Chapter 3

Norms and Factorizations

3.1 Norms of Signals and Systems

In the recent development of control theory, various kinds of norms of signals and systems play important roles. The norm of signals is a non-negative number assigned to each signal which quantifies the "length" of the signal. Some of the norms of signals can induce a norm of systems through the input/output relation generated by the system. The notion of induced norm is the key idea of the contemporary design theory of control systems.

Let $f(t)$ be a signal represented as a time function. The norm of $f(t)$, which is denoted by $\|f\|$, must satisfy the following axioms.

$$(1) \quad \|f\| \geq 0, \ \ \|f\| = 0 \ \text{iff} \ f = 0. \tag{3.1a}$$

$$(2) \quad \|\alpha f\| = |\alpha| \|f\| \quad \text{for each scalar } \alpha. \tag{3.1b}$$

$$(3) \quad \|f_1 + f_2\| \leq \|f_1\| + \|f_2\|. \tag{3.1c}$$

For scalar signal $f(t)$, the following norms are frequently used.

$$\|f\|_1 = \int_0^\infty |f(t)| dt, \tag{3.2}$$

$$\|f\|_2 = \left(\int_0^\infty |f(t)|^2 dt \right)^{1/2}, \tag{3.3}$$

$$\|f\|_\infty = \sup_{t \geq 0} |f(t)|. \tag{3.4}$$

Actually, the preceding three norms are represented as special cases of

$$\|f\|_p = \left(\int_0^\infty |f(t)|^p dt \right)^{1/p}, \quad p > 1. \tag{3.5}$$

If $f(t)$ is a vector-valued signal, that is, $f(t) \in \mathbf{R}^n$, or $f(t) \in \mathbf{C}^n$, we can replace $|f(t)|$ in (3.2)~(3.4) by a suitable norm $\|f(t)\|$ in $\mathbf{R}^n$ or in $\mathbf{C}^n$ to define the norm of $f(t)$. We write

$$\|f\|_1 = \int_0^\infty \|f(t)\| dt, \tag{3.6}$$

$$\|f\|_2 = (\int_0^\infty \|f(t)\|^2 dt)^{1/2}, \tag{3.7}$$

$$\|f\|_\infty = \sup_{t\geq 0} \|f(t)\|, \tag{3.8}$$

$$\|f\|_p = (\int_0^\infty \|f(t)\|^p dt)^{1/p}. \tag{3.9}$$

The norm $\|f(t)\|$ in $\mathbf{R}^n$ or $\mathbf{C}^n$ can be the Euclidean distance

$$\|f(t)\| = (|f_1(t)|^2 + |f_2(t)|^2 + \cdots + |f_n(t)|^2)^{1/2}, \tag{3.10}$$

or the maximum element

$$\|f(t)\| = max(|f_1(t)|,\ |f_2(t)| \cdots |f_n(t)|), \tag{3.11}$$

or anything that satisfies the axioms (3.1) of the norm.

The set of signals that have finite p-norm is denoted by $\mathbf{L}_p$, that is,

$$\mathbf{L}_p = \{f(t) : \|f\|_p < \infty\}. \tag{3.12}$$

In order to explicitly express that the signal is defined for $t \geq 0$, we write

$$\mathbf{L}_p^+ := \{f(t);\ f(t) = 0 \quad (t < 0), \|f\|_p < \infty\}. \tag{3.13}$$

Also, we may write the complement of $\mathbf{L}_p^+$ as

$$\mathbf{L}_p^- := \{f(t);\ f(t) = 0 \quad (t \geq 0), \|f\|_p < \infty\}. \tag{3.14}$$

Each signal $f(t)$ in $\mathbf{L}_p, 1 \leq p < \infty$, must satisfy

$$\|f(t)\| \to 0, \quad t \to \infty, \tag{3.15}$$

because $\|f(t)\|^p$ must be finitely integrable. Therefore, signals of infinite duration such as a unit step or periodic signals such as sinusoidal functions cannot be in $\mathbf{L}_p$. This fact sometimes introduces inconvenience. To circumvent this inconvenience, a new measure of signal, called *power*, is introduced as follows [23]:

$$power(f) = \lim_{T\to\infty} (\frac{1}{T} \int_0^T \|f(t)\|^2 dt)^{1/2}. \tag{3.16}$$

It is easy to see that *power*(f) satisfies (3.1b) and (3.1c), but fails to satisfy (3.1a). Therefore, it is not the norm. However, *power*(f) which expresses the "averaged 2-norm" represents some essential features of signals and is frequently used in many areas of engineering. It is clear that the unit step function has unity power.

Among various norms, the 2-norm (3.7) is of special importance, because it is derived from the inner product in the sense that

$$\|f\|_2 = (< f, f >)^{1/2},$$

where $< f, g >$ denotes the inner product

$$< f, g >= \int_0^\infty f^*(t) g(t) dt. \tag{3.17}$$

In this sense, $\mathbf{L}_2$ is an inner-product space. The space $\mathbf{L}_2$ has another important property. Let $f(s)$ be the Laplace transform of $f(t) \in \mathbf{L}_2^+$, that is,

$$f(s) = \int_0^\infty f(t) e^{-st} dt. \tag{3.18}$$

Then, it is well known that the identity

$$\frac{1}{2\pi} \int_{-\infty}^\infty |f(j\omega)|^2 d\omega = \int_0^\infty |f(t)|^2 dt \tag{3.19}$$

holds, which is usually referred to as the *Parseval's identity*. This implies that $f \in \mathbf{L}_2^+$ always implies $f(j\omega) \in \mathbf{L}_2$, and the norm is preserved to within the factor of $(2\pi)^{-1}$.

Let $s = \sigma + j\omega$ and $\sigma > 0$. Then, from (3.18), it follows that

$$|f(s)| \leq \int_0^\infty |f(t)| e^{-\sigma t} dt < \infty.$$

This implies that $f(s)$ is analytic in the right-half plane. The set of all complex functions $f(s)$ which is analytic in the open *RHP* (*Res* > 0) with $f(j\omega) \in \mathbf{L}_p$ is denoted by $\mathbf{H}_p$, that is,

$$\mathbf{H}_p := \{f(s); \text{analytic in } RHP, \ \ f(j\omega) \in \mathbf{L}_p\}. \tag{3.20}$$

Particularly,

$$\mathbf{H}_\infty =: \{f(s); \text{analytic in } RHP, \ \ \sup_\omega |f(j\omega)| < \infty\}. \tag{3.21}$$

In the rational case, analyticity in the *RHP* implies the stability.

Consider a linear system whose transfer function is given by $G(s)$. The input/output relation of this system is given by

$$y(t) = \int_0^\infty g(t-\tau)u(\tau)d\tau, \tag{3.22}$$

where $g(t)$ is the impulse response of the system computed as the inverse Laplace transform of $G(s)$, that is,

$$g(t) = \mathcal{L}^{-1}[G(s)].$$

Assume that $u(t) \in \mathbf{L}_p$ implies $y(t) \in \mathbf{L}_p$, that is, the system maps $\mathbf{L}_p$ into itself. Then the induced norm of the system is defined as

$$\|G\|_{in} = \sup_{u \in \mathbf{L}_p} \frac{\|y\|_p}{\|u\|_p}. \tag{3.23}$$

It is easy to see that $\|G\|_{in}$ defined above satisfies the axioms of the norm (3.1). Also, it satisfies the submultiplicative property

$$\|G_1G_2\|_{in} \leq \|G_1\|_{in}\|G_2\|_{in}. \tag{3.24}$$

In the case $p = 2$, we can characterize the norm of the system in terms of the transfer function $G(s)$ by using the Laplace transform, that is,

$$\|G\|_{in} = \sup_{u \in L_2} \frac{\|Gu\|_2}{\|u\|_2},$$

where u denotes the Laplace transform of the input $u(t)$. In the SISO case with scalar $G(s)$, it is well-known that the previously induced norm is identical to the $\mathbf{L}_\infty$ norm of $G(s)$, that is,

$$\|G\|_\infty = \sup_\omega |G(j\omega)| = \sup_{u \in L_2} \frac{\|Gu\|_2}{\|u\|_2}. \tag{3.25}$$

In the MIMO case with matrix $G(s)$, the preceding identity becomes

$$\|G\|_\infty = \sup_\omega \bar{\sigma}(G(j\omega)), \tag{3.26}$$

where $\bar{\sigma}(\cdot)$ denotes the maximum singular value. For $p = \infty$. in (3.23), the induced norm is given by

$$\begin{aligned}\|G\|_{in} = \|g\|_1 &= \int_0^\infty |g(t)|dt \\ &= \sup_{u \in L_\infty} \frac{\|y\|_\infty}{\|u\|_\infty}.\end{aligned} \tag{3.27}$$

The above relation is the basis for $\mathbf{L}_1$ control theory which has made a remarkable progress recently.

Example 3.1
Consider a transfer function

$$G(s) = \frac{1}{s+\alpha}, \quad \alpha > 0.$$

Then the infinity norm $\|G\|_\infty$ is calculated to be

$$\|G\|_\infty = \sup_\omega \frac{1}{\sqrt{\omega^2+\alpha^2}} = \frac{1}{\alpha}.$$

Example 3.2
Consider a transfer function

$$G(s) = \frac{s+\beta}{s+\alpha}, \quad \alpha > 0.$$

Then

$$\begin{aligned}\|G\|_\infty &= \sup_\omega \sqrt{\frac{\omega^2+\beta^2}{\omega^2+\alpha^2}} = \left(\sup_\omega (1 - \frac{\alpha^2-\beta^2}{\omega^2+\alpha^2})\right)^{1/2} \\ &= \max\{1, \frac{|\beta|}{\alpha}\}.\end{aligned}$$

Now we exploit the meaning of the infinity norm $\|G\|_\infty$ of the transfer function $G(s)$ from the input/output point of view illustrated in Figure 3.1.

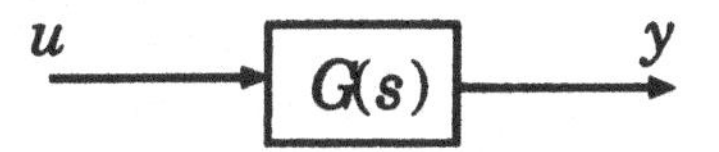

Figure 3.1 Transfer Function $G(s)$.

In terms of Figure 3.1, $\|G\|_\infty \leq \gamma$ implies

(i) the vector locus of $G(s)$ lies within the circle with the origin as its center and the radius γ, as shown in Figure 3.2, provided that $G(s)$ is scalar.

(ii) $\|y\|_2 \leq \gamma \|u\|_2$ for each $u \in L_2$,

(iii) $E[y(t)^2] \leq \gamma^2 E[u(t)^2]$ for each random input with finite variance, where E denotes the expectation,

(iv) $power(y) \leq \gamma\, power(u)$,

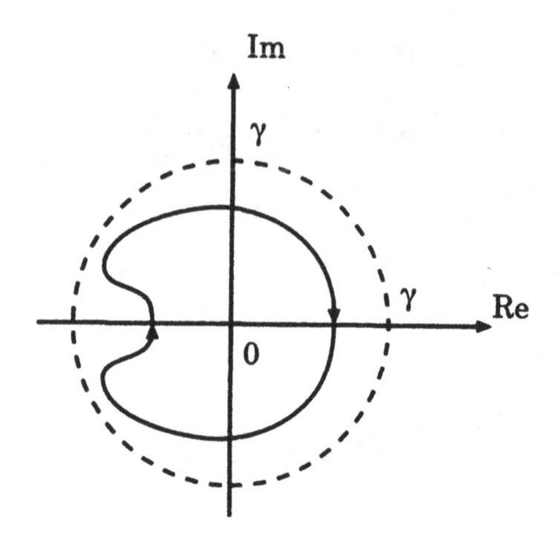

Figure 3.2 Illustration of $\|G\|_\infty \leq \gamma$.

If $G(s)$ is stable and satisfies

$$\|G\|_\infty < 1, \tag{3.28}$$

it is said to be *bounded*. The set of all bounded transfer functions is denoted by $\mathbf{BH}^\infty$, that is,

$$\mathbf{BH}^\infty = \{G(s); G \in \mathbf{H}_\infty, \|G\|_\infty < 1\}. \tag{3.29}$$

It is clear that $G \in \mathbf{BH}^\infty$, iff $G(s)$ is stable and satisfies

$$I - G^\sim(j\omega)G(j\omega) > 0\ , \quad \forall\omega. \tag{3.30}$$

A state-space characterization of bounded functions is given in Section 3.3.

3.2 Hamiltonians and Riccati Equations

A matrix quadratic equation given by

$$XA + A^TX + XWX - Q = 0 \tag{3.31}$$
$$W = W^T \in \mathbf{R}^{n\times n}, \quad Q = Q^T \in \mathbf{R}^{n\times n}$$

is called the *algebraic Riccati equation* (ARE), which plays a fundamental role in linear system theory. Equation (3.31) is associated with a *Hamiltonian* matrix defined by

$$H := \begin{bmatrix} A & W \\ Q & -A^T \end{bmatrix}. \tag{3.32}$$

Simple manipulation verifies the identity

$$\Sigma H + H^T\Sigma = 0, \tag{3.33}$$

where Σ is given by

$$\Sigma = \begin{bmatrix} 0 & -I \\ I & 0 \end{bmatrix}. \tag{3.34}$$

It is easy to see that H is of the form (3.32) iff (3.33) holds. Therefore, the identity (3.33) is a defining property of the Hamiltonian matrix. The Hamiltonian given by (3.32) is closely related to the calculus of variations. We can define the Hamiltonian as a matrix H such that $H\Sigma$ is symmetric, that is,

$$(H\Sigma)^T = H\Sigma. \tag{3.35}$$

Since ARE is a quadratic equation, it has many solutions in $\mathbf{C}^{n\times n}$. Among them, the solution X of (3.31) such that X is real and $A + WX$ is stable is of particular importance. Such a solution is unique. Indeed, let X_1 and X_2 be solutions of (3.31) such that $A + WX_1$ and $A + WX_2$ are stable. Since

$$X_iA + A^TX_i + X_iWX_i - Q = 0, \quad i = 1, 2,$$

we have

$$(X_1 - X_2)(A + WX_1) + (A + WX_2)^T(X_1 - X_2) = 0.$$

Since $A+WX_1$ and $A+WX_2$ are both stable, we conclude that $X_1-X_2 = 0$.

DEFINITION 3.1 *A unique solution X of the ARE (3.31) that stabilizes $A+WX$ is said to be a stabilizing solution of (3.31), and is denoted by*

$$X = Ric(H).$$

Simple manipulations using (3.31) yield

$$\begin{bmatrix} I & 0 \\ -X & I \end{bmatrix} H \begin{bmatrix} I & 0 \\ X & I \end{bmatrix} = \begin{bmatrix} A+WX & W \\ 0 & -(A+WX)^T \end{bmatrix}. \tag{3.36}$$

This identity implies that

$$\sigma(H) = \sigma(A+WX) \cup \sigma(-(A+WX)).$$

If X is a stabilizing solution, $\sigma(A+WX)$ is contained in $\mathbf{C}_-$ and $\sigma(-(A+WX))$ is in $\mathbf{C}_+$. Hence, H does not have eigenvalues on $\mathbf{C}_0$. Therefore, $Ric(H)$ exists only if H has no eigenvalues on the $j\omega$-axis. The converse of the above statement holds under an additional condition on W.

THEOREM 3.2 *Assume that (A, W) is stabilizable and $W \geq 0$. Then $Ric(H)$ exists iff H has no eigenvalues on the imaginary axis.*

Proof. Necessity has already been proven.

To prove the sufficiency, assume that H has no eigenvalues on the $j\omega$-axis. If λ is an eigenvalue of H with x the corresponding eigenvector, we have $(\lambda I + H^T)\Sigma x = 0$ due to (3.33). Hence, $-\lambda$ is an eigenvalue of H. Thus, the eigenvalues of H are distributed symmetrically with respect to the origin in $\mathbf{C}$. Therefore, there exist n eigenvalues of H in the open left-half plane. Denoting the eigenspaces of H corresponding to these stable eigenvalues by $[\ M^T \quad N^T\ ]^T$, we have the relation

$$H \begin{bmatrix} M \\ N \end{bmatrix} = \begin{bmatrix} M \\ N \end{bmatrix} E,$$

where E is a stable matrix in $\mathbf{R}^{n\times n}$. This relation is explicitly represented as

$$AM + WN = ME, \tag{3.37}$$
$$QM - A^T N = NE. \tag{3.38}$$

From (3.38), it follows that

$$N^T AM + N^T WN = N^T ME. \tag{3.39}$$

Since $N^TA = M^TQ - E^TN^T$, we have

$$E^TN^TM + N^TME = M^TQM + N^TWN. \tag{3.40}$$

Since the right-hand side of the identity (3.40) is symmetric, so is its left-hand side. Hence, we have

$$E^T(N^TM - M^TN)^T + (N^TM - M^TN)E = 0.$$

Since E is stable from the assumption, we conclude that $N^TM - M^TN = 0$. Hence, M^TN is symmetric.

Now we show that M is invertible. Assume that $M\xi = 0$ for some ξ. From (3.37) and (3.40), it follows that

$$\xi^TN^TWN\xi = \xi^TN^TME\xi = \xi^TM^TNE\xi = 0.$$

Since $W \geq 0$, we have $WN\xi = 0$. Therefore, the relation (3.37) yields $ME\xi = 0$. This implies that Ker$M$ is E-invariant, in other words, there exists an eigenvalue λ of E such that $E\xi' = \lambda\xi'$, $M\xi' = 0$. The relation (3.38) implies that

$$(\lambda I + A^T)N\xi' = 0, \quad WN\xi' = 0.$$

Since E is stable, Re$\lambda < 0$. From the assumption that (A, W) is stabilizable, we conclude that $N\xi' = 0$. Since $M\xi' = 0$, this contradicts that $[M^T\ N^T]^T$ is column full rank. Therefore, M is invertible.
We finally show that $X := NM^{-1} = Ric(H)$. Due to (3.37) and (3.38),

$$NM^{-1}(AM + WN) = NE = QM - A^TN.$$

Postmultiplication of the preceding relation by M^{-1} easily verifies that $X = NM^{-1}$ satisfies (3.31). Also, from (3.37), it follows that $A + WX = MEM^{-1}$. Since E is stable, so is $A + WX$. Finally, since M^TN is symmetric, $X = NM^{-1} = M^{-T}N^T = X^T$. ∎

From (3.36), $X = Ric(H)$ satisfies

$$H\begin{bmatrix} I \\ X \end{bmatrix} = \begin{bmatrix} I \\ X \end{bmatrix}(A + WX), \quad A + WX : \text{stable}. \tag{3.41}$$

In general, the similarity transformation $T^{-1}HT$ of a Hamiltonian H is no longer a Hamiltonian. However, a special class of transformations preserves the Hamiltonian properties. An example of such transformations is given by

$$T = \begin{bmatrix} I & 0 \\ L & I \end{bmatrix}, \quad L = L^T. \tag{3.42}$$

In that case, the following result holds:

LEMMA 3.3 *Let H_0 be a Hamiltonian matrix with $X_0 = Ric(H_0)$. Then $H = TH_0T^{-1}$ is also a Hamiltonian for any T of the form (3.42) and $X = Ric(H)$ exists which is given by $X = X_0 + L$.*

Proof. Let Σ be given in (3.34). From (3.42), it follows that $T\Sigma T^T = \Sigma$. Since H_0 is a Hamiltonian, we have

$$\begin{aligned} H\Sigma + \Sigma H^T &= -T(H_0T^{-1}\Sigma T^{-T} + T^{-1}\Sigma T^{-T}H_0^T)T^T \\ &= -T(H_0\Sigma + \Sigma H_0^T)T^T = 0. \end{aligned}$$

This implies that H is also a Hamiltonian.

Let $X_0 = Ric(H_0)$. Then it is clear that

$$H_0 \begin{bmatrix} I \\ X_0 \end{bmatrix} = \begin{bmatrix} I \\ X_0 \end{bmatrix} \Lambda$$

for a stable matrix Λ. We have

$$HT \begin{bmatrix} I \\ X_0 \end{bmatrix} = T \begin{bmatrix} I \\ X_0 \end{bmatrix} \Lambda,$$

or, equivalently,

$$H \begin{bmatrix} I \\ X_0 + L \end{bmatrix} = \begin{bmatrix} I \\ X_0 + L \end{bmatrix} \Lambda.$$

This implies that $X_0 + L = Ric(H)$. ∎

The following result is easily shown.

LEMMA 3.4 *If $X = Ric(H)$, then for any nonsingular T,*

$$T^TXT = Ric\left(\begin{bmatrix} T^{-1} & 0 \\ 0 & T^T \end{bmatrix} H \begin{bmatrix} T & 0 \\ 0 & T^{-T} \end{bmatrix} \right). \tag{3.43}$$

Proof. Let $X_1 := T^T X T$. Then from (3.31), it follows that

$$X_1 T^{-1} A T + T^T A^T T^{-T} X_1 + X_1 T^{-1} W T^{-T} X_1 - T^T Q T = 0.$$

Since $A+WX$ is stable, so is $T^{-1}AT + T^{-1}WT^{-T}X_1 = T^{-1}(A+WX)T$. Hence,

$$\begin{aligned} X_1 &= Ric\left(\begin{bmatrix} T^{-1}AT & T^{-1}WT^{-T} \\ T^TQT & -T^TA^TT^{-T} \end{bmatrix}\right) \\ &= Ric\left(\begin{bmatrix} T^{-1} & 0 \\ 0 & T^T \end{bmatrix} H \begin{bmatrix} T & 0 \\ 0 & T^{-T} \end{bmatrix}\right). \end{aligned}$$

∎

The rank defect of the solution X of (3.31) reflects the structural properties of H. The next result is crucial in what follows.

LEMMA 3.5 *Assume that $X = Ric(H)$ is not invertible and $Q \geq 0$. Then any nonsingular matrix T that satisfies*

$$T^T X T = \begin{bmatrix} \bar{X} & 0 \\ 0 & 0 \end{bmatrix}, \quad \bar{X}; \textit{nonsingular} \tag{3.44}$$

transforms each matrix in H as follows,

$$T^{-1} A T = \begin{bmatrix} A_{11} & 0 \\ A_{21} & A_{22} \end{bmatrix}, \quad A_{22}; \textit{stable}, \tag{3.45}$$

$$T^T Q T = \begin{bmatrix} \bar{Q} & 0 \\ 0 & 0 \end{bmatrix}. \tag{3.46}$$

Also, $\bar{X} = Ric(\bar{H})$, where

$$\bar{H} = \begin{bmatrix} A_{11} & W_{11} \\ \bar{Q} & -A_{11}^T \end{bmatrix}, \tag{3.47}$$

$$T^{-1} W T^{-T} = \begin{bmatrix} W_{11} & W_{12} \\ W_{12}^T & W_{22} \end{bmatrix}. \tag{3.48}$$

Proof. Let T be any nonsingular matrix that converts X in the form (3.44). Due to Lemma 3.4,

$$\begin{bmatrix} \bar{X} & 0 \\ 0 & 0 \end{bmatrix} = Ric\left(\begin{bmatrix} T^{-1} & 0 \\ 0 & T^T \end{bmatrix} H \begin{bmatrix} T & 0 \\ 0 & T^{-T} \end{bmatrix}\right). \tag{3.49}$$

Let

$$T^{-1}AT := \begin{bmatrix} A_{11} & A_{12} \\ A_{21} & A_{22} \end{bmatrix}, \quad T^TQT := \begin{bmatrix} \bar{Q} & Q_{12} \\ Q_{12}^T & Q_{22} \end{bmatrix},$$
$$T^{-1}WT^{-T} := \begin{bmatrix} W_{11} & W_{12} \\ W_{12}^T & W_{22} \end{bmatrix}.$$

The size of the partitions is chosen consistently with that of (3.44). The relation (3.49) implies that

$$\begin{bmatrix} \bar{X} & 0 \\ 0 & 0 \end{bmatrix}\begin{bmatrix} A_{11} & A_{12} \\ A_{21} & A_{22} \end{bmatrix} + \begin{bmatrix} A_{11} & A_{12} \\ A_{21} & A_{22} \end{bmatrix}^T\begin{bmatrix} \bar{X} & 0 \\ 0 & 0 \end{bmatrix}$$
$$+ \begin{bmatrix} \bar{X} & 0 \\ 0 & 0 \end{bmatrix}\begin{bmatrix} W_{11} & W_{12} \\ W_{12}^T & W_{22} \end{bmatrix}\begin{bmatrix} \bar{X} & 0 \\ 0 & 0 \end{bmatrix} - \begin{bmatrix} \bar{Q} & Q_{12} \\ Q_{12}^T & Q_{22} \end{bmatrix} = 0 \tag{3.50}$$

and

$$\begin{bmatrix} A_{11} & A_{12} \\ A_{21} & A_{22} \end{bmatrix} + \begin{bmatrix} W_{11} & W_{12} \\ W_{12}^T & W_{22} \end{bmatrix}\begin{bmatrix} \bar{X} & 0 \\ 0 & 0 \end{bmatrix}$$
$$= \begin{bmatrix} A_{11} + W_{11}\bar{X} & A_{12} \\ A_{21} + W_{12}^T\bar{X} & A_{22} \end{bmatrix} \tag{3.51}$$

is stable. It follows, from (3.50), that $Q_{22} = 0$. Since $T^TQT \geq 0$ from the assumption, $Q_{12} = 0$, which implies (3.46). Also, $\bar{X}A_{12} = Q_{12} = 0$ from (3.50). Since $\bar{X}$ is nonsingular, $A_{12} = 0$. Therefore, the stability of the matrix in (3.51) implies that both $A_{11} + W_{11}\bar{X}$ and A_{22} are stable. This implies (3.45). Finally, (3.50) implies $\bar{X}A_{11} + A_{11}^T\bar{X} + \bar{X}W_{11}\bar{X} - \bar{Q} = 0$. This implies $\bar{X} = Ric(\bar{H})$. ∎

The case where $Q = 0$ in (3.31) is of particular interest.

LEMMA 3.6 *The Riccati equation*

$$XA + A^TX + XWX = 0 \tag{3.52}$$

has a stabilizing solution

$$X = Ric(H), \quad H = \begin{bmatrix} A & W \\ 0 & -A^T \end{bmatrix}, \tag{3.53}$$

only if A has no eigenvalue on the imaginary axis. The rank of X is equal to the number of unstable eigenvalues of A and the eigenspace of A corresponding to the stable eigenvalues is equal to Ker$X$.

Proof. Assume that A has an eigenvalue $j\omega$, that is,

$$A\xi = j\omega\xi, \quad \xi \neq 0.$$

Then, due to (3.52), $(j\omega I + A^T + XW)X\xi = 0$. Since $A+WX$ is stable, we conclude that $X\xi = 0$. This implies that $(A+WX)\xi = j\omega\xi$. This contradicts the assumption that $A+WX$ is stable.

Assume that $X = Ric(H)$. If $A\xi = \lambda\xi$ with Re $\lambda \geq 0$, then, from (3.52), it follows that

$$(\lambda I + A^T + XW)X\xi = 0.$$

Since $A^T + XW$ is stable, we conclude that $X\xi = 0$, that is, $\xi \in$ Ker$X$. ∎

Now we show an important result on the monotonicity of Riccati equations that plays a crucial role in what follows.

LEMMA 3.7 *Let $X_1 \geq 0$ and $X_2 \geq 0$ be stabilizing solutions $X_1 = Ric(H_1)$ and $X_2 = Ric(H_2)$, with*

$$H_i = \begin{bmatrix} A_i & W_i \\ Q_i & -A_i^T \end{bmatrix}, \quad i = 1, 2, \tag{3.54}$$

with $Q_2 = 0$. If

$$(H_1 - H_2)\Sigma \geq 0, \tag{3.55}$$

where Σ is given by (3.34), then

$$X_1 \geq X_2. \tag{3.56}$$

The equality in (3.56) holds iff

$$(H_1 - H_2)\begin{bmatrix} I \\ X_1 \end{bmatrix} = 0. \tag{3.57}$$

Proof. We first prove the lemma under the assumption that X_1 is nonsingular. Since $X_1 = Ric(H_1)$, it satisfies

$$[X_1 \ \ -I]H_1\begin{bmatrix} I \\ X_1 \end{bmatrix} = 0$$

or, equivalently,

$$[I \ \ -X_1^{-1}]H_1\begin{bmatrix} X_1^{-1} \\ I \end{bmatrix} = 0.$$

Therefore, from (3.55) and $Q_2 = 0$, it follows that

$$\begin{aligned} X_1^{-1}A_2^T + A_2X_1^{-1} + W_2 &= [I \ \ -X_1^{-1}]H_2\begin{bmatrix} X_1^{-1} \\ I \end{bmatrix} \\ &= -[I \ \ -X_1^{-1}](H_1 - H_2)\begin{bmatrix} X_1^{-1} \\ I \end{bmatrix} \\ &= -[I \ \ -X_1^{-1}](H_1 - H_2)\Sigma\begin{bmatrix} I \\ -X_1^{-1} \end{bmatrix} \le 0. \end{aligned}$$

Therefore, since $X_2A_2 + A_2^TX_2 + X_2W_2X_2 = 0$, we have

$$\begin{aligned} &(X_2X_1^{-1} - I)A_2^TX_2 + X_2A_2(X_1^{-1}X_2 - I) \\ &= X_2(X_1^{-1}A_2^T + A_2X_1^{-1} + W_2)X_2 \le 0. \end{aligned} \tag{3.58}$$

Since $X_2 = Ric(H_2)$ exists, Lemma 3.6 implies that A_2 has no eigenvalues on the imaginary axis. Hence, there exists a transformation T such that

$$T^{-1}A_2T = \begin{bmatrix} A_{2+} & 0 \\ 0 & A_{2-} \end{bmatrix}, \quad A_{2-},\ -A_{2+} \text{ stable.} \tag{3.59}$$

Due to Lemma 3.6, we have

$$T^TX_2T = \begin{bmatrix} \bar{X}_2 & 0 \\ 0 & 0 \end{bmatrix}, \quad \bar{X}_2; \text{nonsingular.} \tag{3.60}$$

Write

$$T^{-1}X_1^{-1}T^{-T} = \begin{bmatrix} \hat{X}_{11} & \hat{X}_{12} \\ \hat{X}_{12}^T & \hat{X}_{22} \end{bmatrix}. \tag{3.61}$$

Substituting (3.59), (3.60), and (3.61) in (3.58) yields

$$(\bar{X}_2\hat{X}_{11} - I)A_{2+}\bar{X}_2 + \bar{X}_2A_{2+}^T(\hat{X}_{11}\bar{X}_2 - I) \leq 0,$$

or, equivalently,

$$(\hat{X}_{11} - \bar{X}_2^{-1})A_{2+} + A_{2+}^T(\hat{X}_{11} - \bar{X}_2^{-1}) \leq 0.$$

Since $-A_{2+}$ is stable, we conclude that $\hat{X}_{11} \leq \bar{X}_2^{-1}$ or, equivalently,

$$\hat{X}_{11}^{-1} \geq \bar{X}_2. \tag{3.62}$$

Write

$$T^TX_1T = (T^{-1}X_1^{-1}T^{-T})^{-1} = \begin{bmatrix} \hat{X}_{11} & \hat{X}_{12} \\ \hat{X}_{12}^T & \hat{X}_{22} \end{bmatrix}^{-1}$$

$$= \begin{bmatrix} X_{11} & X_{12} \\ X_{12}^T & X_{22} \end{bmatrix} > 0.$$

Obviously, $\hat{X}_{11}^{-1} = X_{11} - X_{12}X_{22}^{-1}X_{12}^T$. Now, from (3.60), it follows that

$$T^T(X_1 - X_2)T = \begin{bmatrix} X_{11} - \bar{X}_2 & X_{12} \\ X_{12}^T & X_{22} \end{bmatrix}. \tag{3.63}$$

Since $X_1 > 0$, we have $X_{22} > 0$. The Schur complement of (3.63) is given by $X_{11} - \bar{X}_2 - X_{12}X_{22}^{-1}X_{12}^T = \hat{X}_{11}^{-1} - \bar{X}_2 \geq 0$ due to (3.62). Thus, the lemma has been proven for the case of nonsingular X_1.

Now we show that the general case can be reduced to the non-singular case. Let T be a nonsingular matrix that transforms X_1 to the form

$$T^TX_1T = \begin{bmatrix} \bar{X}_1 & 0 \\ 0 & 0 \end{bmatrix}, \quad \bar{X}_1; \text{nonsingular}.$$

Due to Lemma 3.5,

$$\begin{bmatrix} T^{-1} & 0 \\ 0 & T^T \end{bmatrix} H_1\Sigma \begin{bmatrix} T & 0 \\ 0 & T^{-T} \end{bmatrix} = \begin{bmatrix} W_{11} & W_{11} & A_{11} & 0 \\ W_{12}^T & W_{22} & A_{21} & A_{22} \\ -A_{11}^T & -A_{21}^T & \bar{Q} & 0 \\ 0 & -A_{22}^T & 0 & 0 \end{bmatrix},$$

$$A_{22};\text{stable}, \tag{3.64}$$

where

$$T^{-1}W_1T^{-T} = \begin{bmatrix} W_{11} & W_{12} \\ W_{12}^T & W_{22} \end{bmatrix}, \quad T^{-1}A_1T = \begin{bmatrix} A_{11} & 0 \\ A_{21} & A_{22} \end{bmatrix}.$$

Since

$$\Sigma \begin{bmatrix} T^{-T} & 0 \\ 0 & T \end{bmatrix} = \begin{bmatrix} T & 0 \\ 0 & T^{-T} \end{bmatrix} \Sigma,$$

we have, from (3.55),

$$\begin{aligned} &\begin{bmatrix} T^{-1} & 0 \\ 0 & T^T \end{bmatrix} (H_1 - H_2) \begin{bmatrix} T & 0 \\ 0 & T^{-T} \end{bmatrix} \Sigma \\ &= \begin{bmatrix} T^{-1} & 0 \\ 0 & T^T \end{bmatrix} (H_1 - H_2)\Sigma \begin{bmatrix} T^{-T} & 0 \\ 0 & T \end{bmatrix} \geq 0. \end{aligned} \tag{3.65}$$

Since $Q_2 = 0$, the inequality (3.65) and the identity (3.64) imply that

$$T^{-1}A_2T = \begin{bmatrix} * & 0 \\ * & A_{22} \end{bmatrix}.$$

Since A_2 is stable, the stable eigenspace of $T^{-1}A_2T$ contains $[0 \quad I]^T$. Therefore, due to Lemma 3.6, T^TX_2T is of the form

$$T^TX_2T = \begin{bmatrix} \bar{X}_2 & 0 \\ 0 & 0 \end{bmatrix}.$$

The lemma is proven if $\bar{X}_1 \geq \bar{X}_2$ is shown where

$$\bar{X}_i = Ric(H_i),$$

$$H_i = \begin{bmatrix} \bar{A}_i & \bar{W}_i \\ \bar{Q}_i & -\bar{A}_i^T \end{bmatrix},$$

$$\bar{A}_i = [\ I \quad 0\ ]T^{-1}A_iT\begin{bmatrix} I \\ 0 \end{bmatrix}, \quad \bar{Q}_i = [\ I \quad 0\ ]T^TQ_iT\begin{bmatrix} I \\ 0 \end{bmatrix},$$

$$\bar{W}_i = [\ I \quad 0\ ]T^{-1}W_iT^{-T}\begin{bmatrix} I \\ 0 \end{bmatrix}, \quad i = 1, 2.$$

Obviously, from (3.65), it follows that

$$\begin{bmatrix} \bar{W}_1 & \bar{A}_1 \\ -\bar{A}_1^T & -\bar{Q}_1 \end{bmatrix} - \begin{bmatrix} \bar{W}_2 & \bar{A}_2 \\ -\bar{A}_2^T & 0 \end{bmatrix} \geq 0.$$

Since $\bar{X}_1$ is nonsingular, the proof of the first assertion has been completed. The second assertion is self-evident from the preceding proof.

∎

Example 3.3

In order to illustrate the meaning of Lemma 3.7, consider the scalar case where

$$H_i = \begin{bmatrix} a_i & w_i \\ q_i & -a_i \end{bmatrix}, \quad i = 1, 2, \ q_2 = 0.$$

The associated Riccati equations are the quadratic equations given by

$$f_i(x) = w_ix^2 + 2a_ix - q_i = 0.$$

The condition (3.55) is represented in this case as

$$\begin{bmatrix} w_1 - w_2 & -a_1 + a_2 \\ -a_1 + a_2 & -q_1 \end{bmatrix} \geq 0,$$

which is given by

$$w_1 \geq w_2$$

$$(w_1 - w_2)q_1 + (a_1 - a_2)^2 \leq 0.$$

These inequalities are equivalent to the condition that the quadratic equation

$$f_1(x) = f_2(x)$$

does not have a real solution. Figure 3.3 illustrates the situation when $w_2 > 0$.

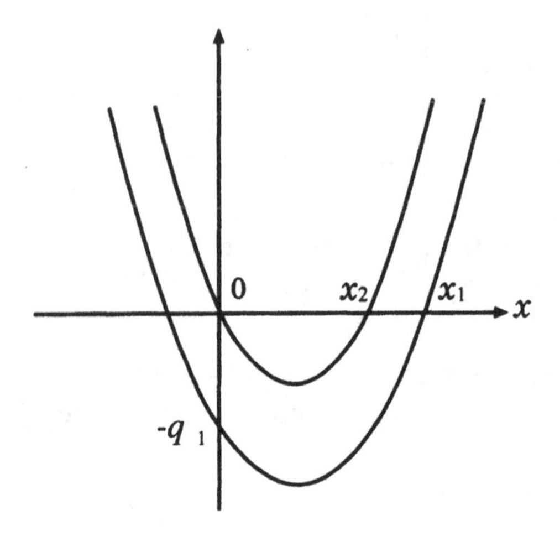

Figure 3.3 Illustration of Lemma 3.6.

3.3 Factorizations

In linear system theory, we encounter various kinds of factorizations such as the Wiener-Hopf factorization of positive matrices, the inner/outer factorization, the coprime factorization, and so on. Among them, the most important one is the Wiener-Hopf factorization of positive matrices, in which Riccati equations discussed in the preceding section play fundamental roles.

Let $\Gamma(s)$ be a rational Hermite matrix on the $j\omega$-axis, that is,

$$\Gamma(s) = \Gamma^{\sim}(s). \tag{3.66}$$

Assume also that $\Gamma(s)$ has no poles on the imaginary axis. Taking the partial fraction expansion of $\Gamma(s)$ that breaks it into the stable part $\Phi(s)$ and the anti-stable part $\Phi^{\sim}(s)$, that is,

$$\Gamma(s) = \Phi(s) + \Phi^{\sim}(s), \tag{3.67}$$

assume that $\Phi(s)$ has a minimal state-space realization

$$\Phi(s) = \left[\begin{array}{c|c} A & B \\ \hline C & (1/2)R \end{array}\right]. \quad A : \textit{stable}. \tag{3.68}$$

Then (3.67) implies that $\Gamma(s)$ has a realization

$$\Gamma(s) = \left[\begin{array}{cc|c} A & 0 & B \\ 0 & -A^T & C^T \\ \hline C & -B^T & R \end{array}\right]. \tag{3.69}$$

If R is invertible, $\Gamma(s)$ has a rational inverse $\Gamma(s)^{-1}$ which has a realization

$$\Gamma(s)^{-1} = \left[\begin{array}{cc|c} A - BR^{-1}C & BR^{-1}B^T & -B \\ -C^TR^{-1}C & -A^T + C^TR^{-1}B^T & C^T \\ \hline R^{-1}C & R^{-1}B^T & I \end{array}\right] R^{-1}. \tag{3.70}$$

Let us denote by H the A-matrix of the realization (3.70), that is,

$$H := \left[\begin{array}{cc} A - BR^{-1}C & BR^{-1}B^T \\ -C^TR^{-1}C & -A^T + C^TR^{-1}B^T \end{array}\right]. \tag{3.71}$$

It is easily seen that H is a Hamiltonian in the sense that it satisfies (3.33). The following theorem is one of the most important results in linear system theory.

THEOREM 3.8 *A rational matrix $\Gamma(s)$ given by (3.69) is positive, that is,*

$$\Gamma(j\omega) > 0, \quad \forall\omega, \tag{3.72}$$

iff $R = R^T > 0$ and the Riccati equation associated with H in (3.71) has a stabilizing solution $X = Ric(H)$. In that case, $\Gamma(s)$ is represented as

$$\Gamma(s) = M^{\sim}(s)RM(s), \tag{3.73}$$

where

$$M(s) = \left[\begin{array}{c|c} A & B \\ \hline -F & I \end{array}\right], \tag{3.74}$$

$$F = -R^{-1}(C - B^TX), \tag{3.75}$$

is unimodular.

Proof. If (3.72) holds, $\Gamma(s)$ has no zeros on the imaginary axis. Hence, $\Gamma(s)^{-1}$ has no poles on the $j\omega$-axis, which implies that H has no eigenvalues on the $j\omega$-axis. Therefore, since $\Gamma(\infty) = R > 0$, Theorem 3.2 implies the existence of $X = Ric(H)$.

Conversely, assume that $X = Ric(H)$ exists for H given by (3.71). We show that $M(s)$ given by (3.74) satisfies (3.73). Due to (3.31), $X = Ric(H)$ satisfies the Riccati equation

$$X(A - BR^{-1}C) + (A - BR^{-1}C)^T X + XBR^{-1}B^T X + C^T R^{-1} C = 0, \tag{3.76}$$

which is rewritten as

$$XA + A^T X + F^T RF = 0, \tag{3.77}$$

where F is given by (3.75). Since X is a stabilizing solution of (3.76), $A + BF$ is stable. Direct manipulations yield

$$M^{\sim}(s)RM(s) = \left[\begin{array}{cc|c} -A^T & -F^T RF & F^T R \\ 0 & A & B \\ \hline B^T & -RF & R \end{array}\right].$$

Taking a similarity transformation of the above state-space form with the transformation matrix

$$T = \begin{bmatrix} 0 & I \\ I & -X \end{bmatrix}, \quad T^{-1} = \begin{bmatrix} X & I \\ I & 0 \end{bmatrix},$$

and using (3.77) and $B^T X - RF = C$, we can easily verify (3.73). Since $R > 0$, we have (3.72).

Finally, from

$$M(s)^{-1} = \left[\begin{array}{c|c} A + BF & B \\ \hline F & I \end{array}\right],$$

it follows that $M(s)$ is unimodular.

As an application of the preceding theorem, we can derive the so-called *Bounded Real Lemma*. Assume that a stable transfer function $G(s)$ has a state-space realization

$$G(s) = \left[\begin{array}{c|c} A & B \\ \hline C & D \end{array}\right]. \tag{3.78}$$

The concatenation rule (2.13) yields

$$G^{\sim}(s)G(s) = \left[\begin{array}{cc|c} A & 0 & B \\ -C^TC & -A^T & -C^TD \\ \hline D^TC & B^T & D^TD \end{array}\right] \tag{3.79}$$

Let W be the observability gramian of (3.78), that is,

$$WA + A^TW + C^TC = 0. \tag{3.80}$$

The similarity transformation of (3.79) with

$$T = \begin{bmatrix} I & 0 \\ -W & I \end{bmatrix}, \quad T^{-1} = \begin{bmatrix} I & 0 \\ W & I \end{bmatrix}$$

yields

$$G^{\sim}(s)G(s) = \left[\begin{array}{cc|c} A & 0 & B \\ 0 & -A^T & -(WB + C^TD) \\ \hline B^TW + D^TC & B & D^TD \end{array}\right]. \tag{3.81}$$

The identity (3.81) can be written as

$$\begin{aligned} G^{\sim}(s)G(s) &= \Phi(s) + \Phi^{\sim}(s) + D^TD \\ \Phi(s) &:= \left[\begin{array}{c|c} A & B \\ \hline B^TW + D^TC & 0 \end{array}\right], \end{aligned} \tag{3.82}$$

which represents the partial fraction decomposition of $G^{\sim}(s)G(s)$ into the stable part $\Phi(s)$ and the anti-stable part $\Phi^{\sim}(s)$.

Now we are ready to state the Bounded Real Lemma.

THEOREM 3.9 (Bounded Real Lemma) *Assume that a stable transfer function $G(s)$ has a state-space form (3.78). Then the following statements are equivalent.*

(i) $\|G\|_\infty < 1$. $(G \in \mathbf{BH}^\infty)$.

(ii) $R := I - D^TD > 0$ *and* $X = Ric(H)$ *exist, where*

$$H := \begin{bmatrix} A + BR^{-1}D^TC & BR^{-1}B^T \\ -C^T(I - DD^T)^{-1}C & -(A + BR^{-1}D^TC)^T \end{bmatrix}; \tag{3.83}$$

that is, there exists a solution $X \geq 0$ to the Riccati equation

$$XA + A^TX + (XB + C^TD)R^{-1}(B^TX + D^TC) + C^TC = 0$$

which stabilizes

$$\hat{A} := A + BR^{-1}(B^TX + D^TC).$$

Proof. The assertion (i) is equivalent to

$$\Gamma(s) := I - G^{\sim}(s)G(s) > 0, \quad \forall s = j\omega.$$

Due to the expression (3.81), $\Gamma(s)$ has a state-space form

$$\Gamma(s) = \left[\begin{array}{cc|c} A & 0 & -B \\ 0 & -A^T & WB + C^TD \\ \hline B^TW + D^TC & B & R \end{array}\right]. \tag{3.84}$$

Clearly, $R = I - D^TD = \Gamma(\infty) > 0$. The state-space form (3.84) corresponds to (3.69) with the B and C being replaced by $-B$ and $D^TC + B^TW$, respectively. Hence, Theorem 3.8 implies the existence of $X_0 = Ric(H_0)$ with

$$H_0 := \begin{bmatrix} A + BR^{-1}(D^TC + B^TW) & BR^{-1}B^T \\ -(C^TD + WB)R^{-1}(D^TC + B^TW) & -A^T - (C^TD + WB)R^{-1}B^T \end{bmatrix}.$$

Some manipulation using (3.80) verifies the identity

$$H = \begin{bmatrix} I & 0 \\ W & I \end{bmatrix} H_0 \begin{bmatrix} I & 0 \\ -W & I \end{bmatrix}.$$

Therefore, taking $L = W$ in Lemma 3.3 verifies that $X := X_0 + W = Ric(H)$. ∎

Remark: The Hamiltonian H given by (3.83) is written as

$$H = \begin{bmatrix} A & 0 \\ -C^TC & -A^T \end{bmatrix} - \begin{bmatrix} -B \\ C^TD \end{bmatrix} R^{-1} \begin{bmatrix} D^TC & B^T \end{bmatrix}.$$

Using (3.79), we can see that H is identical to the A-matrix of $(I - G^{\sim}(s)G(s))^{-1}$. If $R > 0, X = Ric(H)$ exists, iff H has no eigenvalues on the $j\omega$-axis due to Theorem 3.2. Therefore, assertion (ii) of Theorem 3.9 is equivalent to $R > 0$ and H has no eigenvalues on the $j\omega$-axis. Thus, we have the following restatement of Theorem 3.9.

COROLLARY 3.10 *The following statements are equivalent.*

(i) $\|G\|_\infty < 1 \quad (G \in \mathbf{BH}^\infty)$.

(ii) G *is stable,* $I - G^{\sim}(\infty)G(\infty) > 0$, *and* $(I - G^{\sim}(s)G(s))^{-1}$ *has no poles on the* $j\omega$*-axis.*

Example 3.4
Consider a first order system

$$G(s) = \frac{1}{s+\alpha}, \quad \alpha > 0$$

which was treated in Example 3.1. The corresponding Hamiltonian matrix in this case is calculated to be

$$H = \begin{bmatrix} -\alpha & 1 \\ -1 & \alpha \end{bmatrix}.$$

It is clear that H has no eigenvalues on the $j\omega$-axis iff $\alpha > 1$. This coincides with the computation in Example 3.1.

The following result is an immediate consequence of Theorem 3.8.

COROLLARY 3.11 *Assume that* $G(s)$ *is a stable transfer function with state-space form (3.78). The following statements are equivalent.*

(i) $\|G\|_\infty < \gamma$.

(ii) $R_\gamma := \gamma^2 I - D^T D > 0$ *and* $X = Ric(H_\gamma)$

exist, where

$$H_\gamma := \begin{bmatrix} A + BR_\gamma^{-1}D^T C & BR_\gamma^{-1}B^T \\ -\gamma^2 C^T(\gamma^2 I - DD^T)^{-1}C & (A + BR_\gamma^{-1}D^T C)^T \end{bmatrix};$$

in other words, there exists a solution X to the Riccati equation

$$XA + A^TX + (XB + C^TD)R_\gamma^{-1}(XB + C^TD)^T + C^TC = 0 \tag{3.85}$$

which stabilizes

$$\hat{A} := A + BR_\gamma^{-1}(XB + C^TD)^T. \tag{3.86}$$

Proof. The condition (i) is equivalent to

$$\|\gamma^{-1}G\|_\infty < 1,$$

where $\gamma^{-1}G(s)$ has a realization

$$\gamma^{-1}G(s) = \left[\begin{array}{c|c} A & \gamma^{-1}B \\ \hline C & \gamma^{-1}D \end{array}\right]. \tag{3.87}$$

Application of Theorem 3.4 to (3.87) proves the assertion.

Now we state another well-known factorization. If a stable matrix $\Sigma(s)$ satisfies

$$\Sigma^\sim(s)\Sigma(s) = I, \quad \forall s, \tag{3.88}$$

then $\Sigma(s)$ is said to be *lossless* or *inner.* If a stable transfer function $Q(s)$ is invertible with stable inverse, then $Q(s)$ is said to be *unimodular* or *outer.* If a stable transfer function is represented as

$$G(s) = \Sigma(s)Q(s), \tag{3.89}$$

where $\Sigma(s)$ is inner and $Q(s)$ is outer, the representation (3.89) is said to be an *inner-outer factorization* of $G(s)$. Due to (3.88), we have

$$G^\sim(s)G(s) = Q^\sim(s)Q(s). \tag{3.90}$$

Conversely, if there exists an outer $Q(s)$ satisfying (3.90), $\Sigma(s) := G(s)Q(s)^{-1}$ satisfies (3.88). Since $\Sigma(s)$ is obviously stable, $\Sigma(s)$ is inner. Thus, $G(s)$ has an inner-outer factorization, if and only if there exists an outer $Q(s)$ satisfying (3.90). Actually, almost all stable transfer functions with left inverse have an inner-outer factorization.

THEOREM 3.12 *Assume that a stable transfer function $G(s)$ has a state-space realization (3.78). Then $G(s)$ has an inner-outer factorization (3.89), iff $G^\sim(s)G(s)$ has no zeros on the imaginary axis and*

$$R := D^TD > 0. \tag{3.91}$$

In that case, factors in (3.89) are given, respectively, by

$$\Sigma(s) = \left[\begin{array}{c|c} A+BF & B \\ \hline C+DF & D \end{array}\right] R^{-1/2}, \tag{3.92}$$

$$Q(s) = R^{1/2} \left[\begin{array}{c|c} A & B \\ \hline -F & I \end{array}\right], \tag{3.93}$$

where

$$F = -R^{-1}(D^T C - B^T X) \tag{3.94}$$

and $X = Ric(H)$ with

$$H = \begin{bmatrix} A - BR^{-1}D^T C & BR^{-1}B^T \\ C^T UC & -(A - BR^{-1}D^T C)^T \end{bmatrix}, \tag{3.95}$$

$$U = I - D(D^T D)^{-1} D^T. \tag{3.96}$$

In other words, X is a solution to the Riccati equation

$$XA + A^T X + (XB - C^T D)R^{-1}(B^T X - D^T C) - C^T C = 0 \tag{3.97}$$

that stabilizes $A + BF$.

Proof. If $G(s)$ has an inner-outer factorization (3.89), then, due to (3.90), $R := D^T D = G^\sim(\infty)G(\infty) = Q^\sim(\infty)Q(\infty)$. Since $Q(s)$ is outer, $Q(\infty)$ must be nonsingular. Therefore, $R > 0$. Note that H is an A-matrix of $(G^\sim(s)G(s))^{-1}$ that has no poles on the $j\omega$-axis from the assumption. Since $U \geq 0$, $X = Ric(H)$ exists according to Theorem 3.2. Direct manipulation verifies that the factors (3.92) and (3.93) satisfy (3.89). Since $X = Ric(H)$, $A + BF$ is stable, and so is

$$Q(s)^{-1} = \left[\begin{array}{c|c} A+BF & B \\ \hline F & I \end{array}\right] R^{-1/2}.$$

Thus, $Q(s)$ is outer.

If $G(s)$ itself is lossless, then $G(s) = \Sigma(s)$ and $Q(s) = I$ in (3.89). This implies that $F = 0$ and $R = I$ in Theorem 3.12. Thus $XB = C^T D$. Putting $P = -X$ in Theorem 3.12, we have the following well-known characterization of lossless systems.

THEOREM 3.13 *A transfer function $G(s)$ whose realization is given by (3.78) is lossless iff there exists a matrix $P \geq 0$ such that*

$$PA + A^T P + C^T C = 0, \tag{3.98a}$$
$$PB + C^T D = 0, \tag{3.98b}$$
$$D^T D = I. \tag{3.98c}$$

Sometimes, we encounter a stable transfer function $G(s)$ such that

$$G(s)G^{\sim}(s) = I. \tag{3.99}$$

Such a system is called *dual lossless* or *co-inner*. A state-space characterization of dual lossless systems is easily obtained by dualizing Theorem 3.13.

THEOREM 3.14 *A transfer function $G(s)$ whose realization is given by* (3.78) *is dual lossless iff there exists a matrix $Q \geq 0$ such that*

$$QA^T + AQ + BB^T = 0, \tag{3.100a}$$
$$CQ + DB^T = 0, \tag{3.100b}$$
$$DD^T = I. \tag{3.100c}$$

We conclude the section by stating an application of the bounded real lemma (Theorem 3.8) to the proof of the *Small Gain Theorem*, which is concerned with the unity feedback system depicted in Figure 3.4.

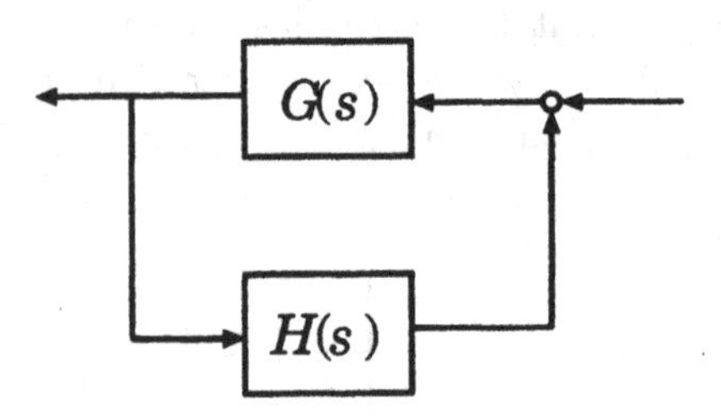

Figure 3.4 The Unity Feedback System.

THEOREM 3.15 *The closed-loop system of Figure 3.4 is stable if both $G(s)$ and $H(s)$ are stable and satisfy*

$$\|G(s)\|_\infty < \gamma, \quad \|H(s)\|_\infty < \gamma^{-1}. \tag{3.101}$$

Proof. We can assume, without loss of generality, that $\gamma = 1$. Also, we can assume that $G(s)$ is strictly proper; that is, $G(\infty) = 0$. Let

$$G(s) = \left[\begin{array}{c|c} A & B \\ \hline C & 0 \end{array}\right], \quad H(s) = \left[\begin{array}{c|c} F & G \\ \hline H & D \end{array}\right]$$

be minimal realizations of $G(s)$ and $H(s)$. Due to Theorem 3.8, the conditions (3.97) imply that there exist $P \geq 0$ and $Q \geq 0$ such that

$$PA + A^T P + PBB^T P + C^T C = 0,$$
$$QF + F^T Q + (QG + H^T D)R^{-1}(G^T Q + D^T H) + H^T H = 0,$$

where $R = I - D^T D > 0$. The closed-loop system of Figure 3.4 has a state-space realization with the A-matrix being given by

$$\hat{A} := \begin{bmatrix} A + BDC & BH \\ GC & F \end{bmatrix}.$$

Straightforward manipulations yield the identity

$$\begin{bmatrix} P & 0 \\ 0 & Q \end{bmatrix}\hat{A} + \hat{A}^T\begin{bmatrix} P & 0 \\ 0 & Q \end{bmatrix} + \begin{bmatrix} C^T D^T - PB \\ H^T \end{bmatrix}\begin{bmatrix} DC - B^T P & H \end{bmatrix}$$
$$+ \begin{bmatrix} C^T R \\ -(QG + H^T D) \end{bmatrix} R^{-1} \begin{bmatrix} RC & -(G^T Q + D^T H) \end{bmatrix} = 0.$$

It is not difficult to see that the pair

$$\left(\hat{A}, \begin{bmatrix} DC - B^T P & H \\ RC & -(G^T Q + D^T H) \end{bmatrix}\right) \tag{3.102}$$

is detectable (See Problem 3.5). Due to Lemma 2.14, $\hat{A}$ is stable. ∎

Notes

Almost all the materials in this chapter are found in standard textbooks of state-space theory such as [11] except Lemma 3.7, which is new and has not been published elsewhere. It is a different version of the well-known monotonicity result due to Wimmer [101] which was stated as Problem 3.3. The Riccati equation has been the fertile resource of many theoretical issues in modern control theory. A compilation of the recent results on Riccati equations can be found in the monograph [10][65]. The results in Section 3.3 are all standard, but the derivations are nonstandard with emphasis on clarity and simplicity.

Problems

[1] Assume that H_0 is a Hamiltonian matrix and let

$$T = \begin{bmatrix} T_{11} & T_{12} \\ T_{21} & T_{22} \end{bmatrix}$$

satisfy the identity

$$T^T \Sigma T = \Sigma ,$$

where Σ is a matrix given by (3.34). Show that $H = TH_0T^{-1}$ is a Hamiltonian matrix. Prove that if $X_0 = Ric(H_0)$ exists, then

$$X := (T_{11} + T_{12}X_0)^{-1}(T_{21} + T_{22}X_0) = Ric(H). \qquad (3.103)$$

[2] Assume that a stable system $G(s)$ given by

$$G(s) = \left[\begin{array}{c|c} A & B \\ \hline C & 0 \end{array}\right]$$

satisfies $\|G\|_\infty < 1$. Show that $A + BC$ and $A - BC$ are both stable.

[3] Let X_1 and X_2 be stabilizing solutions $X_1 = Ric(H_1)$ and $X_2 = Ric(H_2)$ with

$$H_i = \begin{bmatrix} A_i & W_i \\ Q_i & -A_i^T \end{bmatrix}, \quad W_i \geq 0, \quad i = 1, 2.$$

Prove that $(H_1 - H_2)\Sigma \geq 0$ implies $X_1 \geq X_2$, where Σ is given by (3.34).

[4] Compute an inner/outer factorization of the system

$$G(s) = \left[\begin{array}{cc|c} 0 & 1 & 0 \\ -2 & -3 & 1 \\ \hline -10 & -2 & 2 \end{array}\right] = \frac{2(s-1)(s+3)}{(s+1)(s+2)}.$$

[5] Show that the pair (3.102) is detectable.

[6] Assume that $Q \leq 0$, $W \geq 0$ in (3.31), (A, Q) is detectable, and (A, W) is stabilizable. Show that A is stable iff (3.31) has a solution $X \geq 0$.

[7] Assume that $W \leq 0$ in (3.31), (A, W) is stabilizable, and $X_0 = Ric(H)$ exists. Show that X_0 is the maximum solution of (3.31) in the sense that it satisfies

$$X_0 \geq X$$

for any solution of (3.31).

[8] Let

$$\Phi(t) = e^{Ht} = \begin{bmatrix} \Phi_{11}(t) & \Phi_{12}(t) \\ \Phi_{21}(t) & \Phi_{22}(t) \end{bmatrix},$$

where H is a Hamiltonian matrix given in (3.32), and assume that $\Phi_{11}(t)$ is invertible. Show that $X(t) = \Phi_{21}(t)\Phi_{11}(t)^{-1}$ satisfies the Riccati differential equation

$$-\dot{X} = XA + A^TX + XWX - Q.$$

[9] Show that $\|C(sI - A)^{-1}B\|_\infty < \gamma\delta$, iff

$$XA + A^TX + \frac{1}{\gamma^2}XBB^TX + \frac{1}{\delta^2}C^TC \leq 0$$

for some X.

Chapter 4

Chain-Scattering Representations of the Plant

4.1 Algebra of Chain-Scattering Representation

Consider a system Σ of Figure 4.1 with two kinds of inputs (b_1, b_2) and two kinds of outputs (a_1, a_2) represented as

$$\begin{bmatrix} a_1 \\ a_2 \end{bmatrix} = \Sigma \begin{bmatrix} b_1 \\ b_2 \end{bmatrix} = \begin{bmatrix} \Sigma_{11} & \Sigma_{12} \\ \Sigma_{21} & \Sigma_{22} \end{bmatrix} \begin{bmatrix} b_1 \\ b_2 \end{bmatrix}. \tag{4.1}$$

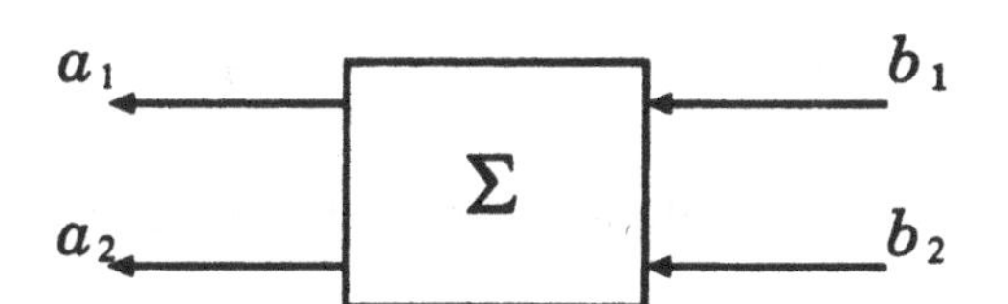

Figure 4.1 Input-Output Representation.

Assuming that Σ_{21} is invertible, we have

$$b_1 = \Sigma_{21}^{-1}(a_2 - \Sigma_{22} b_2).$$

Substituting this relation in the first equation of (4.1) yields

$$a_1 = (\Sigma_{12} - \Sigma_{11}\Sigma_{21}^{-1}\Sigma_{22})b_2 + \Sigma_{11}\Sigma_{21}^{-1}a_2.$$

Therefore, if we write

$$CHAIN\,(\Sigma) := \begin{bmatrix} \Sigma_{12} - \Sigma_{11}\Sigma_{21}^{-1}\Sigma_{22} & \Sigma_{11}\Sigma_{21}^{-1} \\ -\Sigma_{21}^{-1}\Sigma_{22} & \Sigma_{21}^{-1} \end{bmatrix}, \tag{4.2}$$

the relation (4.1) is alternatively represented as

$$\begin{bmatrix} a_1 \\ b_1 \end{bmatrix} = CHAIN\,(\Sigma) \begin{bmatrix} b_2 \\ a_2 \end{bmatrix}. \tag{4.3}$$

The representation (4.3) of the plant is usually referred to as a *chain-scattering representation* of Σ in classical circuit theory. If Σ represents a usual input/output relation of a system, $CHAIN\,(\Sigma)$ represents a *port characteristic* that more or less reflects the physical structure of the system. The chain-scattering representation describes the system as a *wave scatterer* between the (b_2, a_1)-wave and the (b_1, a_2)-wave that travel opposite each other (Figure 4.2).

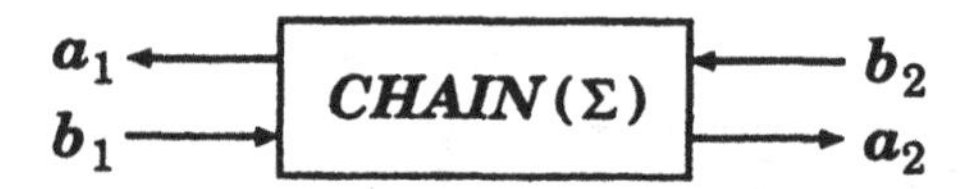

Figure 4.2 Chain-Scattering Representation.

The main reason for using the chain-scattering representation lies in its simplicity of representing cascade connection. The cascade connection of two chain-scattering representations Θ_1 and Θ_2, which actually contains feedback connections of the two systems, is represented simply as the product $\Theta_1\Theta_2$ of each chain-scattering matrix, as is shown in Figure 4.3.

The same connection is represented in the input/output format of (4.1) in Figure 4.4. The resulting expression is much more complicated and is called the *star-product* by Redheffer [81].

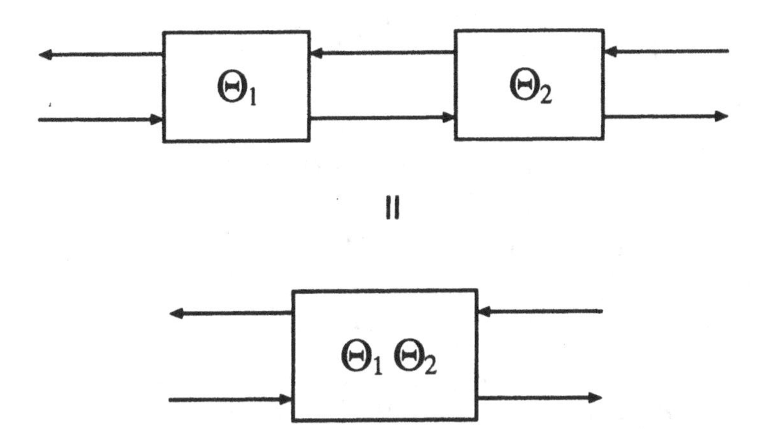

Figure 4.3 Cascade Property of Chain-Scattering Representation.

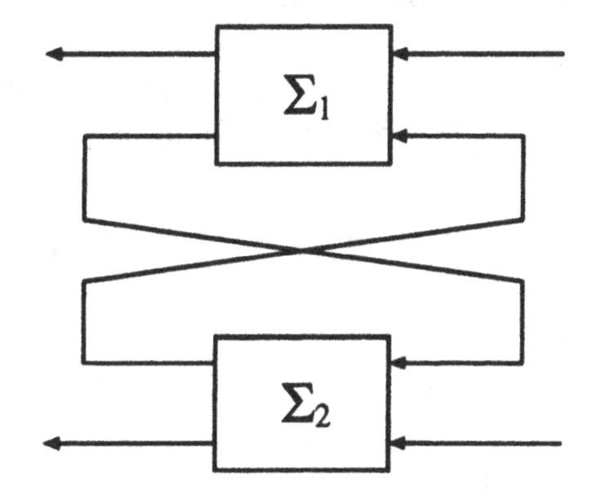

Figure 4.4 Star-Product.

The transformation (4.2) is represented simply as

$$CHAIN\,(\Sigma) = \begin{bmatrix} \Sigma_{12} & \Sigma_{11} \\ 0 & I \end{bmatrix} \begin{bmatrix} I & 0 \\ \Sigma_{22} & \Sigma_{21} \end{bmatrix}^{-1}. \tag{4.4}$$

If Σ is stable, the representation (4.4) can be regarded as a right coprime factorization of $CHAIN\,(\Sigma)$. To see that it is coprime, let U be a common right denominator of the factors of (4.4); that is,

$$\begin{bmatrix} \Sigma_{12} & \Sigma_{11} \\ 0 & I \end{bmatrix} = NU, \quad \begin{bmatrix} I & 0 \\ \Sigma_{22} & \Sigma_{21} \end{bmatrix} = MU.$$

Obviously,

$$\begin{bmatrix} M_{11} & M_{12} \\ N_{21} & N_{22} \end{bmatrix} = U^{-1},$$

which implies that U is unimodular. Therefore, (4.4) is a coprime factorization.

The transformation (4.2) is also represented as

$$CHAIN\,(\Sigma) = \begin{bmatrix} I & \Sigma_{11} \\ 0 & -\Sigma_{21} \end{bmatrix}^{-1} \begin{bmatrix} \Sigma_{12} & 0 \\ -\Sigma_{22} & I \end{bmatrix}, \tag{4.5}$$

which is regarded as a left coprime factorization of $CHAIN\,(\Sigma)$ when Σ is stable.

It is interesting to derive alternative descriptions of the coprime factorizations (4.4) and (4.5). Let us define the following "partial systems,"

$$\Sigma_{1\cdot} := \begin{bmatrix} \Sigma_{11} & \Sigma_{12} \\ I & 0 \end{bmatrix}, \quad \Sigma_{2\cdot} := \begin{bmatrix} 0 & I \\ \Sigma_{21} & \Sigma_{22} \end{bmatrix}. \tag{4.6}$$

It is easy to see that

$$CHAIN\,(\Sigma_{1\cdot}) = \begin{bmatrix} \Sigma_{12} & \Sigma_{11} \\ 0 & I \end{bmatrix},$$

$$CHAIN\,(\Sigma_{2\cdot}) = \begin{bmatrix} I & 0 \\ \Sigma_{22} & \Sigma_{21} \end{bmatrix}^{-1}.$$

Therefore, it follows, from (4.4), that

$$CHAIN\,(\Sigma) = CHAIN\,(\Sigma_{1\cdot}) \cdot CHAIN\,(\Sigma_{2\cdot}). \tag{4.7}$$

In a similar way, we can represent (4.5) as

$$CHAIN\,(\Sigma) = CHAIN\,(\Sigma_{\cdot 1}) \cdot CHAIN\,(\Sigma_{\cdot 2}), \tag{4.8}$$

where

$$\Sigma_{\cdot 1} := \begin{bmatrix} \Sigma_{11} & I \\ \Sigma_{21} & 0 \end{bmatrix}, \quad \Sigma_{\cdot 2} := \begin{bmatrix} 0 & \Sigma_{12} \\ I & \Sigma_{22} \end{bmatrix}. \tag{4.9}$$

The meaning of the representation (4.7) is illustrated in Figure 4.5, where $\Sigma_{1\cdot}$ and $\Sigma_{2\cdot}$ are identified with Σ_1 and Σ_2 in Figure 4.4 , respectively. Thus, Σ is represented as the star product of $\Sigma_{1\cdot}$ and $\Sigma_{2\cdot}$ implying (4.7).

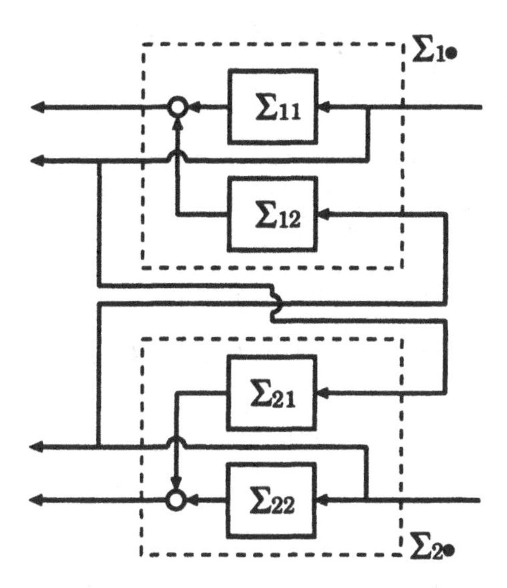

Figure 4.5 Illustration of (4.7).

An analogous reasoning can be obtained for the representation (4.8), which is left to the reader.

Let

$$\Theta = \begin{bmatrix} \Theta_{11} & \Theta_{12} \\ \Theta_{21} & \Theta_{22} \end{bmatrix} = CHAIN\,(\Sigma).$$

Then Σ is represented in terms of Θ as

$$\Sigma = \begin{bmatrix} \Theta_{12}\Theta_{22}^{-1} & \Theta_{11} - \Theta_{12}\Theta_{22}^{-1}\Theta_{21} \\ \Theta_{22}^{-1} & -\Theta_{22}^{-1}\Theta_{21} \end{bmatrix}. \tag{4.10}$$

Note that Θ_{22} is always invertible if Θ is a chain-scattering matrix. The representation (4.10) has factored forms

$$\Sigma = \begin{bmatrix} \Theta_{11} & \Theta_{12} \\ 0 & I \end{bmatrix} \begin{bmatrix} \Theta_{21} & \Theta_{22} \\ I & 0 \end{bmatrix}^{-1} \tag{4.11}$$

$$= \begin{bmatrix} I & -\Theta_{12} \\ 0 & \Theta_{22} \end{bmatrix}^{-1} \begin{bmatrix} 0 & \Theta_{11} \\ I & -\Theta_{21} \end{bmatrix}. \tag{4.12}$$

The transformation (4.2) is also represented in terms of coprime factorizations of the input/output form (4.1). Let Σ in (4.1) be represented as a right coprime factorization

$$\Sigma = \begin{bmatrix} N_{11} & N_{12} \\ N_{21} & N_{22} \end{bmatrix} \begin{bmatrix} M_{11} & M_{12} \\ M_{21} & M_{22} \end{bmatrix}^{-1}, \tag{4.13}$$

where the sizes of the block matrices in (4.13) are conformable to the partition of Σ. Then, by introducing an intermediate variable $c = [c_1^T \; c_2^T]^T$, the relation (4.13) is represented as

$$\begin{bmatrix} a_1 \\ a_2 \end{bmatrix} = \begin{bmatrix} N_{11} & N_{12} \\ N_{21} & N_{22} \end{bmatrix} \begin{bmatrix} c_1 \\ c_2 \end{bmatrix},$$

$$\begin{bmatrix} b_1 \\ b_2 \end{bmatrix} = \begin{bmatrix} M_{11} & M_{12} \\ M_{21} & M_{22} \end{bmatrix} \begin{bmatrix} c_1 \\ c_2 \end{bmatrix}.$$

It follows that

$$\begin{bmatrix} a_1 \\ b_1 \end{bmatrix} = \begin{bmatrix} N_{11} & N_{12} \\ M_{11} & M_{12} \end{bmatrix} \begin{bmatrix} c_1 \\ c_2 \end{bmatrix}, \quad \begin{bmatrix} b_2 \\ a_2 \end{bmatrix} = \begin{bmatrix} M_{21} & M_{22} \\ N_{21} & N_{22} \end{bmatrix} \begin{bmatrix} c_1 \\ c_2 \end{bmatrix}.$$

Thus, we have a representation of $CHAIN\,(\Sigma)$ as

$$CHAIN\,(\Sigma) = \begin{bmatrix} N_{11} & N_{12} \\ M_{11} & M_{12} \end{bmatrix} \begin{bmatrix} M_{21} & M_{22} \\ N_{21} & N_{22} \end{bmatrix}^{-1}. \tag{4.14}$$

If Σ is represented by a left coprime factorization

$$\Sigma = \begin{bmatrix} \hat{M}_{11} & \hat{M}_{12} \\ \hat{M}_{21} & \hat{M}_{22} \end{bmatrix}^{-1} \begin{bmatrix} \hat{N}_{11} & \hat{N}_{12} \\ \hat{N}_{21} & \hat{N}_{22} \end{bmatrix}, \tag{4.15}$$

then we have

$$\begin{bmatrix} \hat{M}_{11} & -\hat{N}_{11} \\ \hat{M}_{21} & -\hat{N}_{21} \end{bmatrix} \begin{bmatrix} a_1 \\ b_1 \end{bmatrix} = \begin{bmatrix} \hat{N}_{12} & -\hat{M}_{12} \\ \hat{N}_{22} & -\hat{M}_{22} \end{bmatrix} \begin{bmatrix} b_2 \\ a_2 \end{bmatrix}.$$

This yields a representation of $CHAIN\,(\Sigma)$ as

$$CHAIN\,(\Sigma) = \begin{bmatrix} \hat{M}_{11} & -\hat{N}_{11} \\ \hat{M}_{21} & -\hat{N}_{21} \end{bmatrix}^{-1} \begin{bmatrix} \hat{N}_{12} & -\hat{M}_{12} \\ \hat{N}_{22} & -\hat{M}_{22} \end{bmatrix}. \tag{4.16}$$

Although the chain-scattering representation is not widely known in the control community, it is a popular way of representing physical systems from the viewpoint of power port analysis.

Example 4.1 Electrical Circuit
An electrical circuit with two ports in Figure 4.6 is represented by a cascade matrix F as

$$\begin{bmatrix} v_1 \\ i_1 \end{bmatrix} = F \begin{bmatrix} v_2 \\ i_2 \end{bmatrix}.$$

This is a typical example of chain-scattering representation. The F-matrix is used for filter design. It is a chain-scattering representation of a hybrid matrix H defined as

$$\begin{bmatrix} v_1 \\ i_2 \end{bmatrix} = H \begin{bmatrix} i_1 \\ v_2 \end{bmatrix}$$

$$F = CHAIN\,(H).$$

The hybrid representation is convenient for representing a circuit containing transistors.

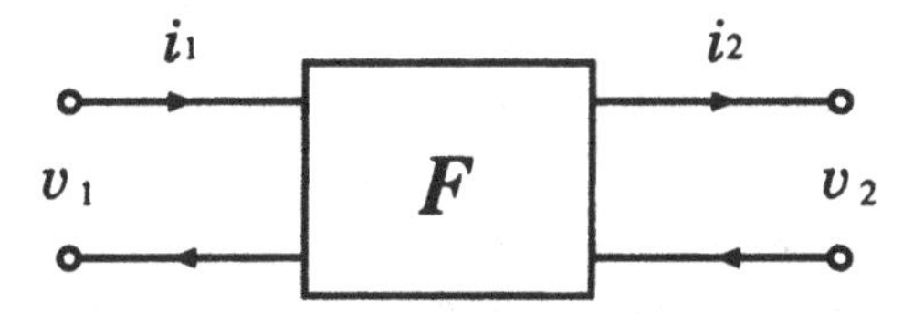

Figure 4.6 Electrical Circuit.

Example 4.2 Acoustic Tube
Consider an acoustic tube as shown in Figure 4.7. The pressure $p(x,t)$ and the particle velocity $v(x,t)$ are related as

$$\frac{\partial p}{\partial x} + \rho \frac{\partial v}{\partial t} = 0,$$

$$\frac{\partial p}{\partial t} + \kappa \frac{\partial v}{\partial x} = 0,$$

where ρ and κ are the density and the elasticity of the air, respectively. Now it is clear that both p and v satisfy the wave equations

$$\frac{\partial^2 p}{\partial t^2} = c^2 \frac{\partial^2 p}{\partial x^2}, \quad \frac{\partial^2 v}{\partial t^2} = c^2 \frac{\partial^2 v}{\partial x^2},$$

$$c^2 = \frac{\kappa}{\rho}.$$

The relation between the pair of port variables p and v at $x = 0$ and $x = L$ for sinusoidal behaviors is described by

$$\begin{bmatrix} p \\ v \end{bmatrix}_{x=L} = U \begin{bmatrix} p \\ v \end{bmatrix}_{x=0},$$

$$U = \begin{bmatrix} \cosh(\gamma Ls) & \sinh(\gamma Ls) \\ \sinh(\gamma Ls) & \cosh(\gamma Ls) \end{bmatrix},$$

where $\gamma := \sqrt{\rho/\kappa} = 1/c$. Obviously, U is a chain-scattering representation.

Figure 4.7 An Acoustic Tube.

Example 4.3 Vibration Suppression
Consider a mechanical system of Figure 4.8. The mass m_1 connected to a fixed surface by a spring k_1 and a dashpot c_1 receives an external force w. The active force u is applied to m_1 and m_2 to suppress the passive vibration absorber composed of a spring k_2 and a dashpot c_2. Taking x_1 and x_2 as displacements of m_1 and m_2, respectively, we have the following equations of motion,

$$m_1\ddot{x}_1 + c_1\dot{x}_1 + c_2(\dot{x}_1 - \dot{x}_2) + k_1 x_1 + k_2(x_1 - x_2) = w - u,$$
$$m_2\ddot{x}_2 + c_2(\dot{x}_2 - \dot{x}_1) + k_2(x_2 - x_1) = u.$$

If we choose

$$z = \begin{bmatrix} \dot{x}_1 \\ u \end{bmatrix}$$

as the controlled output and

$$y = m_0\ddot{x}_1 + c_0\dot{x}_1 + k_0 x_1$$

as the observed output, we have the plant description:

$$\begin{bmatrix} z \\ y \end{bmatrix} = \Sigma \begin{bmatrix} w \\ u \end{bmatrix}$$

$$\Sigma = \left[\begin{array}{c|c} \dfrac{sq(s)}{l(s)} & -\dfrac{m_2 s^3}{l(s)} \\ 0 & 1 \\ \hline \dfrac{q(s)p(s)}{l(s)} & -\dfrac{m_2 s^2 p(s)}{l(s)} \end{array}\right],$$

where

$$\begin{aligned} l(s) &= (m_1 s^2 + c_1 s + k_1)(m_2 s^2 + c_2 s + k_2) + m_2 s^2 (c_2 s + k_2), \\ q(s) &= m_2 s^2 + c_2 s + k_2, \\ p(s) &= m_0 s^2 + c_0 s + k_0. \end{aligned}$$

According to (4.2), we have

$$CHAIN\,(\Sigma) = \left[\begin{array}{c|c} 0 & \dfrac{s}{p(s)} \\ 1 & 0 \\ \hline \dfrac{m_2 s^2}{q(s)} & \dfrac{l(s)}{p(s)q(s)} \end{array}\right],$$

which is slightly simpler than the original expression of Σ. Now, from (4.6), it follows that

$$\Sigma_{1\cdot} = \left[\begin{array}{c|c} \dfrac{sq(s)}{l(s)} & -\dfrac{m_2 s^3}{l(s)} \\ 0 & 1 \\ \hline 1 & 0 \end{array}\right],$$

$$\Sigma_{2\cdot} = \left[\begin{array}{c|c} 0 & 1 \\ \hline \dfrac{q(s)p(s)}{l(s)} & -\dfrac{m_2 s^2 p(s)}{l(s)} \end{array}\right].$$

According to (4.2), we have

$$
CHAIN\,(\Sigma_{1\cdot}) = \left[\begin{array}{c|c} -\dfrac{m_2 s^3}{l(s)} & \dfrac{sq(s)}{l(s)} \\ 1 & 0 \\ \hline 0 & 1 \end{array}\right],
$$

$$
CHAIN\,(\Sigma_{2\cdot}) = \left[\begin{array}{c|c} 1 & 0 \\ \hline \dfrac{m_2 s^2}{q(s)} & \dfrac{l(s)}{p(s)q(s)} \end{array}\right].
$$

It is straightforward to see that (4.7) holds.

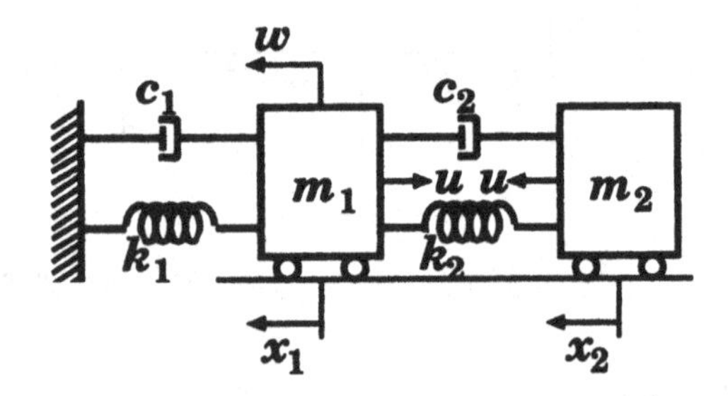

Figure 4.8 A Mechanical System.

4.2 State-Space Forms of Chain-Scattering Representation

Assume that the system Σ in (4.1) has a state-space realization

$$
\Sigma = \begin{bmatrix} \Sigma_{11} & \Sigma_{12} \\ \Sigma_{21} & \Sigma_{22} \end{bmatrix} = \left[\begin{array}{c|cc} A & B_1 & B_2 \\ \hline C_1 & D_{11} & D_{12} \\ C_2 & D_{21} & D_{22} \end{array}\right]. \tag{4.17}
$$

If Σ_{21}^{-1} exists and is proper, the chain-scattering representation (4.2) is also a proper rational matrix and can be represented in a state-space form. Now, according to (4.17), a state-space form of Σ in (4.1) is described as

$$
\dot{x} \;=\; Ax + B_1 b_1 + B_2 b_2, \tag{4.18a}
$$

$$a_1 = C_1x + D_{11}b_1 + D_{12}b_2, \tag{4.18b}$$
$$a_2 = C_2x + D_{21}b_1 + D_{22}b_2. \tag{4.18c}$$

It is clear that Σ_{21}^{-1} exists and is proper if and only if D_{21}^{-1} exists. In that case, the relation (4.18c) is written as

$$b_1 = D_{21}^{-1}(a_2 - C_2x - D_{22}b_2).$$

Substituting this relation in (4.18a) and (4.18b) yields

$$\dot{x} = (A - B_1D_{21}^{-1}C_2)x + (B_2 - B_1D_{21}^{-1}D_{22})b_2 + B_1D_{21}^{-1}a_2,$$
$$a_1 = (C_1 - D_{11}D_{21}^{-1}C_2)x + (D_{12} - D_{11}D_{21}^{-1}D_{22})b_2 + D_{11}D_{21}^{-1}a_2.$$

Thus, we have the following state-space realization of $CHAIN\,(\Sigma)$,

$$CHAIN\,(\Sigma) = \left[\begin{array}{c|cc} A - B_1D_{21}^{-1}C_2 & B_2 - B_1D_{21}^{-1}D_{22} & B_1D_{21}^{-1} \\ \hline C_1 - D_{11}D_{21}^{-1}C_2 & D_{12} - D_{11}D_{21}^{-1}D_{22} & D_{11}D_{21}^{-1} \\ -D_{21}^{-1}C_2 & -D_{21}^{-1}D_{22} & D_{21}^{-1} \end{array}\right]. \tag{4.19}$$

An interesting feature of the state-space form (4.19) is its connection with state-space forms of the factored representations (4.4) and (4.5). Due to (4.17), state-space realizations of $\Sigma_{1\cdot}$ and $\Sigma_{2\cdot}$ in (4.6) are given by

$$\Sigma_{1\cdot} = \begin{bmatrix} \Sigma_{11} & \Sigma_{12} \\ I & 0 \end{bmatrix} = \left[\begin{array}{c|cc} A & B_1 & B_2 \\ \hline C_1 & D_{11} & D_{12} \\ 0 & I & 0 \end{array}\right] = \left[\begin{array}{c|c} A & B \\ \hline \hat{C}_1 & D_{1\cdot} \end{array}\right], \tag{4.20}$$

$$\Sigma_{2\cdot} = \begin{bmatrix} 0 & I \\ \Sigma_{21} & \Sigma_{22} \end{bmatrix} = \left[\begin{array}{c|cc} A & B_1 & B_2 \\ \hline 0 & 0 & I \\ C_2 & D_{21} & D_{22} \end{array}\right] = \left[\begin{array}{c|c} A & B \\ \hline \hat{C}_2 & D_{2\cdot} \end{array}\right], \tag{4.21}$$

where

$$B := \begin{bmatrix} B_1 & B_2 \end{bmatrix},\ \hat{C}_1 := \begin{bmatrix} C_1 \\ 0 \end{bmatrix},\ \hat{C}_2 := \begin{bmatrix} 0 \\ C_2 \end{bmatrix},$$
$$D_{1\cdot} := \begin{bmatrix} D_{11} & D_{12} \\ I & 0 \end{bmatrix},\ D_{2\cdot} := \begin{bmatrix} 0 & I \\ D_{21} & D_{22} \end{bmatrix}. \tag{4.22}$$

A state-space form (4.19) can be represented in a compact way in terms of the state-space forms of $\Sigma_{1\cdot}$ and $\Sigma_{2\cdot}$ as follows.

LEMMA 4.1 *A state-space form (4.19) of CHAIN* (Σ) *is represented as*

$$CHAIN\,(\Sigma) = \left[\begin{array}{c|c} A - BD_{2\cdot}^{-1}\hat{C}_2 & B \\ \hline \hat{C}_1 - D_{1\cdot}D_{2\cdot}^{-1}\hat{C}_2 & D_{1\cdot} \end{array}\right] D_{2\cdot}^{-1}. \tag{4.23}$$

Proof. Noting that

$$D_{2\cdot}^{-1} = \begin{bmatrix} -D_{21}^{-1}D_{22} & D_{21}^{-1} \\ I & 0 \end{bmatrix},$$

we can prove the equivalence of (4.19) and (4.23) immediately. ∎

The representation (4.23) implies that *CHAIN* (Σ) is generated by applying a *state feedback* $K = -D_{2\cdot}^{-1}\hat{C}_2$ and an *input transformation* $U = D_{2\cdot}^{-1}$ to the system $\Sigma_{1\cdot}$ given by (4.20). The representation (4.23) corresponds to the factored representation (4.7). The state-space forms (4.17) and (4.19) also establish the following important result.

LEMMA 4.2 *If the state-space realization (4.17) of* Σ *is minimal, then so is the corresponding state-space form (4.19).*

Proof. If (A, B) is controllable, so is the state-space form (4.19). Analogous arguments apply to observability. ∎

Let

$$m = \dim(a_1),\ r = \dim(a_2) = \dim(b_1),\ p = \dim(b_2).$$

Introducing the notation

$$J_{mr} := \begin{bmatrix} I_m & 0 \\ 0 & -I_r \end{bmatrix}, \tag{4.24}$$

we can easily establish the following identities.

$$C^T C = \hat{C}_1^T J_{mr}\hat{C}_1 - \hat{C}_2^T J_{pr}\hat{C}_2, \tag{4.25a}$$

$$C^T D = \hat{C}_1^T J_{mr} D_{1\cdot} - \hat{C}_2^T J_{pr} D_{2\cdot}, \tag{4.25b}$$

$$I_{p+r} - D^T D = D_{2\cdot}^T J_{pr} D_{2\cdot} - D_{1\cdot}^T J_{mr} D_{1\cdot}, \tag{4.25c}$$

where

$$C = \begin{bmatrix} C_1 \\ C_2 \end{bmatrix},\quad D = \begin{bmatrix} D_{11} & D_{12} \\ D_{21} & D_{22} \end{bmatrix}.$$

These relations are used later.

4.3 Dualization

If Σ_{12} is invertible in (4.1), we can solve the first equation of (4.1) with respect to b_2 to obtain

$$b_2 = \Sigma_{12}^{-1}(a_1 - \Sigma_{11}b_1). \tag{4.26}$$

Substituting this relation in the second relation of (4.1), we have an alternative representation of (4.1) as

$$\begin{bmatrix} b_2 \\ a_2 \end{bmatrix} = \begin{bmatrix} \Sigma_{12}^{-1} & -\Sigma_{12}^{-1}\Sigma_{11} \\ \Sigma_{22}\Sigma_{12}^{-1} & \Sigma_{21} - \Sigma_{22}\Sigma_{12}^{-1}\Sigma_{11} \end{bmatrix} \begin{bmatrix} a_1 \\ b_1 \end{bmatrix}. \tag{4.27}$$

We write

$$DCHAIN\,(\Sigma) := \begin{bmatrix} \Sigma_{12}^{-1} & -\Sigma_{12}^{-1}\Sigma_{11} \\ \Sigma_{22}\Sigma_{12}^{-1} & \Sigma_{21} - \Sigma_{22}\Sigma_{12}^{-1}\Sigma_{11} \end{bmatrix}, \tag{4.28}$$

which is called a *dual chain-scattering representation* of Σ. If both Σ_{21} and Σ_{12} are invertible; that is, if both $CHAIN\,(\Sigma)$ and $DCHAIN\,(\Sigma)$ exist, they are inversely related, that is,

$$CHAIN\,(\Sigma) \cdot DCHAIN\,(\Sigma) = I. \tag{4.29}$$

The duality between the primal and its inverse is another salient characteristic feature of the chain-scattering representation.

It is easy to see that $DCHAIN\,(\Sigma)$ in (4.28) is also represented as

$$DCHAIN\,(\Sigma) = \begin{bmatrix} I & 0 \\ \Sigma_{22} & \Sigma_{21} \end{bmatrix} \begin{bmatrix} \Sigma_{12} & \Sigma_{11} \\ 0 & I \end{bmatrix}^{-1} \tag{4.30}$$

$$= \begin{bmatrix} \Sigma_{12} & 0 \\ -\Sigma_{22} & I \end{bmatrix}^{-1} \begin{bmatrix} I & -\Sigma_{11} \\ 0 & \Sigma_{21} \end{bmatrix}. \tag{4.31}$$

These representations correspond to (4.4) and (4.5), respectively. Dual to (4.7) and (4.8), these representations are rewritten as

$$DCHAIN\,(\Sigma) = DCHAIN\,(\Sigma_{2\cdot})DCHAIN\,(\Sigma_{1\cdot}) \tag{4.32}$$

$$= DCHAIN\,(\Sigma_{\cdot 2})DCHAIN\,(\Sigma_{\cdot 1}), \tag{4.33}$$

respectively, where $\Sigma_{\cdot i}$ and $\Sigma_{i\cdot}$ $(i = 1, 2)$ are given in (4.6) and (4.9), respectively.

If

$$\Psi = \begin{bmatrix} \Psi_{11} & \Psi_{12} \\ \Psi_{21} & \Psi_{22} \end{bmatrix} = DCHAIN\,(\Sigma),$$

then we can represent Σ in terms of Ψ as

$$\Sigma = \begin{bmatrix} -\Psi_{11}^{-1}\Psi_{12} & \Psi_{11}^{-1} \\ \Psi_{22} - \Psi_{21}\Psi_{11}^{-1}\Psi_{12} & \Psi_{21}\Psi_{11}^{-1} \end{bmatrix}.$$

This actually represents the inverse transformation of *DCHAIN*, which is also represented as

$$\Sigma = \begin{bmatrix} I & 0 \\ \Psi_{21} & \Psi_{22} \end{bmatrix} \begin{bmatrix} 0 & I \\ \Psi_{11} & \Psi_{12} \end{bmatrix}^{-1} \tag{4.34}$$

$$= \begin{bmatrix} \Psi_{11} & 0 \\ -\Psi_{21} & I \end{bmatrix}^{-1} \begin{bmatrix} -\Psi_{12} & I \\ \Psi_{22} & 0 \end{bmatrix}. \tag{4.35}$$

Associated with the right coprime factorization (4.13) of Σ, *DCHAIN* (Σ) can be represented as

$$DCHAIN\,(\Sigma) = \begin{bmatrix} M_{21} & M_{22} \\ N_{21} & N_{22} \end{bmatrix} \begin{bmatrix} N_{11} & N_{12} \\ M_{11} & M_{12} \end{bmatrix}^{-1}. \tag{4.36}$$

The representation of *DCHAIN* (Σ) corresponding to the left coprime factorization (4.15) is given by

$$DCHAIN\,(\Sigma) = \begin{bmatrix} \hat{N}_{12} & -\hat{M}_{12} \\ \hat{N}_{22} & -\hat{M}_{22} \end{bmatrix}^{-1} \begin{bmatrix} \hat{M}_{11} & -\hat{N}_{11} \\ \hat{M}_{21} & -\hat{N}_{21} \end{bmatrix}. \tag{4.37}$$

If Σ_{12}^{-1} is proper, we can obtain a state-space realization of the *DCHAIN* (Σ) in terms of a state-space realization (4.17) of Σ. Since Σ_{12}^{-1} is proper if and only if D_{12}^{-1} exists, we have, from (4.18b),

$$b_2 = D_{12}^{-1}(a_1 - C_1 x - D_{11} b_1).$$

Substitution of this relation in (4.18a) yields

$$\dot{x} = (A - B_2 D_{12}^{-1} C_1)x + B_2 D_{12}^{-1} a_1 + (B_1 - B_2 D_{12}^{-1} D_{11}) b_1.$$

This, together with (4.18c), gives a realization of $DCHAIN\,(\Sigma)$ as

$$DCHAIN\,(\Sigma) = \left[\begin{array}{c|cc} A - B_2 D_{12}^{-1} C_1 & B_2 D_{12}^{-1} & B_1 - B_2 D_{12}^{-1} D_{11} \\ \hline -D_{12}^{-1} C_1 & D_{12}^{-1} & -D_{12}^{-1} D_{11} \\ C_2 - D_{22} D_{12}^{-1} C_1 & D_{22} D_{12}^{-1} & D_{21} - D_{22} D_{12}^{-1} D_{11} \end{array}\right]. \tag{4.38}$$

The dual version of Lemma 4.1 is obtained in terms of state-space realization of the factors in (4.31) which are given by

$$\hat{\Sigma}_{\cdot 1} := \Sigma_{\cdot 1} \begin{bmatrix} 0 & I \\ -I & 0 \end{bmatrix} = \begin{bmatrix} -I & \Sigma_{11} \\ 0 & \Sigma_{21} \end{bmatrix} = \left[\begin{array}{c|c} A & \hat{B}_1 \\ \hline C & D_{\cdot 1} \end{array}\right], \tag{4.39}$$

$$\hat{\Sigma}_{\cdot 2} := \Sigma_{\cdot 2} \begin{bmatrix} 0 & I \\ -I & 0 \end{bmatrix} = \begin{bmatrix} -\Sigma_{12} & 0 \\ -\Sigma_{22} & I \end{bmatrix} = \left[\begin{array}{c|c} A & \hat{B}_2 \\ \hline C & D_{\cdot 2} \end{array}\right], \tag{4.40}$$

where

$$\begin{gathered} C = \begin{bmatrix} C_1 \\ C_2 \end{bmatrix}, \quad \hat{B}_1 = \begin{bmatrix} 0 & B_1 \end{bmatrix}, \quad \hat{B}_2 := \begin{bmatrix} -B_2 & 0 \end{bmatrix}, \\ D_{\cdot 1} := \begin{bmatrix} -I & D_{11} \\ 0 & D_{21} \end{bmatrix}, \quad D_{\cdot 2} := \begin{bmatrix} -D_{12} & 0 \\ -D_{22} & I \end{bmatrix}. \end{gathered} \tag{4.41}$$

LEMMA 4.3 *A state-space form of $DCHAIN\,(\Sigma)$ is represented as*

$$DCHAIN\,(\Sigma) = D_{\cdot 2}^{-1} \left[\begin{array}{c|c} A - \hat{B}_2 D_{\cdot 2}^{-1} C & \hat{B}_1 - \hat{B}_2 D_{\cdot 2}^{-1} D_{\cdot 1} \\ \hline C & D_{\cdot 1} \end{array}\right]. \tag{4.42}$$

The state-space realization (4.42) implies that $DCHAIN\,(\Sigma)$ is obtained by performing an *output insertion* $L = -\hat{B}_2 D_{\cdot 2}^{-1}$ and an *output transformation* $V = D_{\cdot 2}^{-1}$ to the system $\hat{\Sigma}_{\cdot 1}$ given by (4.39). If we identify the sizes of signals in (4.1) as

$$m = \dim(a_1) = \dim(b_2), \quad q = \dim(a_2), \quad r = \dim(b_1), \tag{4.43}$$

we have the identities:

$$BB^T = \hat{B}_2 J_{mq} \hat{B}_2^T - \hat{B}_1 J_{mr} \hat{B}_{1\cdot}^T, \tag{4.44}$$

$$BD^T = \hat{B}_2 J_{mq} D_{\cdot 2}^T - \hat{B}_1 J_{mr} D_{\cdot 1}^T, \tag{4.45}$$

$$I_{m+q} - DD^T = D_{\cdot 1} J_{mr} D_{\cdot 1}^T - D_{\cdot 2} J_{mq} D_{\cdot 2}^T, \tag{4.46}$$

where

$$B = \begin{bmatrix} B_1 & B_2 \end{bmatrix}, \quad D = \begin{bmatrix} D_{11} & D_{12} \\ D_{21} & D_{22} \end{bmatrix}.$$

In (4.43), the equality $\dim(a_1) = \dim(b_2)$ comes from the assumption that Σ_{12} is invertible.

4.4 J-Lossless and (J, J')-Lossless Systems

A system Σ is said to be *lossless* if Σ is stable and the input power is always equal to the output power; that is,

$$\|b_1(j\omega)\|^2 + \|b_2(j\omega)\|^2 = \|a_1(j\omega)\|^2 + \|a_2(j\omega)\|^2 \tag{4.47}$$

for each ω. An equivalent representation of (4.47) is given by

$$\Sigma^{\sim}(s)\Sigma(s) = I \ , \ \forall s, \tag{4.48}$$

which is equivalent to (??). If the stability is not required, Σ is said to be *unitary*.

Assume that Σ has a chain-scattering representation

$$\Theta = CHAIN\,(\Sigma).$$

Preservation of power expressed in (4.47) is written in terms of the port variables of (4.3) as

$$\|a_1(j\omega)\|^2 - \|b_1(j\omega)\|^2 = \|b_2(j\omega)\|^2 - \|a_2(j\omega)\|^2, \tag{4.49}$$

which is equivalently represented as

$$\Theta^{\sim}(s)J_{mr}\Theta(s) = J_{pr}, \tag{4.50}$$

where $m = \dim(a_1)$, $r = \dim(a_2) = \dim(b_1)$, $p = \dim(b_2)$, and J_{mr} denotes the signature matrix

$$J_{mr} = \begin{bmatrix} I_m & 0 \\ 0 & -I_r \end{bmatrix}. \tag{4.51}$$

A matrix Θ satisfying (4.48) is called (J_{mr}, J_{pr})-*unitary*. If $m = p$, then it is simply called J_{mr}-*unitary*.

Another important property of lossless systems is the inequality

$$\Sigma^*(s)\Sigma(s) \leq I , \ \forall \text{Re } s \geq 0, \tag{4.52}$$

which comes from the stability of $\Sigma(s)$ and the Maximum Modulus Theorem [2]. This property is passed on to its chain-scattering representation as

$$\Theta^*(s)J_{mr}\Theta(s) \leq J_{pr} , \ \forall \text{Re } s \geq 0, \tag{4.53}$$

as we show in the next lemma.

A matrix $\Theta(s)$ satisfying (4.50) and (4.53) is called (J_{mr}, J_{pr})-*lossless*. We sometimes use the abbreviated notation (J, J')-*lossless* if the sizes of J_{mr} and J_{pr} are clear from the context or irrelevant.

The preceding argument implies that if Σ is lossless and $\Theta = CHAIN\ (\Sigma)$ exists, then Θ is (J_{mr}, J_{pr})-lossless. Actually, a much stronger result holds in the following lemma.

LEMMA 4.4 *A matrix $\Theta(s)$ is (J_{mr}, J_{pr})-unitary (lossless) iff it is a chain-scattering representation of a unitary (lossless) matrix.*

Proof. Let

$$\Theta = \begin{bmatrix} \Theta_{11} & \Theta_{12} \\ \Theta_{21} & \Theta_{22} \end{bmatrix}$$

be a partition of Θ with Θ_{11} and Θ_{22} being $m \times p$ and $r \times r$, respectively. Due to (4.50), we have

$$\Theta_{12}^{\sim}\Theta_{12} - \Theta_{22}^{\sim}\Theta_{22} = -I_r , \ \forall s.$$

This implies that Θ_{22}^{-1} exists. Let

$$N := \begin{bmatrix} \Theta_{11} & \Theta_{12} \\ 0 & I \end{bmatrix}, \ M := \begin{bmatrix} \Theta_{21} & \Theta_{22} \\ I & 0 \end{bmatrix},$$

and define

$$\Sigma = NM^{-1}.$$

This corresponds to (4.11), and hence $\Theta = CHAIN\ (\Sigma)$. It follows that

$$M^{\sim}(s)M(s) - N^{\sim}(s)N(s) = J_{pr} - \Theta^{\sim}(s)J_{mr}\Theta(s).$$

The assertion is now obvious, because Σ is unitary iff $M^{\sim}M = N^{\sim}N$ and Σ is lossless iff it is unitary and $M^*M \geq N^*N$ for each Re $s \geq 0$. ∎

A state-space characterization of (J, J')-lossless systems is well known. We give a proof based on the state-space characterization (4.19) of the chain-scattering representation. Let

$$\Theta(s) = \left[\begin{array}{c|c} A_\theta & B_\theta \\ \hline C_\theta & D_\theta \end{array}\right] \in \mathbf{RL}^\infty_{(m+r)\times(p+r)} \tag{4.54}$$

be a state-space realization of Θ.

THEOREM 4.5 *A matrix $\Theta(s)$ given by (4.54) is (J_{mr}, J_{pr})-unitary, if and only if*

$$D_\theta^T J_{mr} D_\theta = J_{pr} \tag{4.55}$$

and there exists a matrix P satisfying

$$PA_\theta + A_\theta^T P + C_\theta^T J_{mr} C_\theta = 0, \tag{4.56a}$$

$$D_\theta^T J_{mr} C_\theta + B_\theta^T P = 0. \tag{4.56b}$$

The solution P satisfies $P \geq 0$ iff Θ is (J_{mr}, J_{pr})-lossless. The solution satisfies $P > 0$ iff Θ is (J_{mr}, J_{pr})-lossless and (A_θ, C_θ) is observable.

Proof. Due to Lemma 4.4, there exists a unitary system Σ such that $\Theta = CHAIN\ (\Sigma)$. Let

$$\Sigma = \left[\begin{array}{c|c} A & B \\ \hline C & D \end{array}\right]$$

be a realization of Σ. Theorem 3.13 implies that $D^T D = I$ and there exists a matrix P such that

$$PA + A^T P + C^T C = 0, \ PB + C^T D = 0. \tag{4.57}$$

Due to Lemma 4.1,

$$A_\theta = A - BD_{2\cdot}^{-1}\hat{C}_2, \ B_\theta = BD_{2\cdot}^{-1}, \tag{4.58}$$

$$C_\theta = \hat{C}_1 - D_{1\cdot}D_{2\cdot}^{-1}\hat{C}_2, \ D_\theta = D_{1\cdot}D_{2\cdot}^{-1}, \tag{4.59}$$

where $\hat{C}_2, D_{2\cdot}, \hat{C}_1, D_{1\cdot}$ are given by (4.22). Write

$$J = J_{mr}, \ J' = J_{pr}$$

for simplicity. Since $D^T D = I$, we have, from (4.25c),

$$D_{2\cdot}^T J' D_{2\cdot} = D_{1\cdot}^T J D_{1\cdot}, \tag{4.60}$$

which implies (4.55) due to (4.59). Also, from (4.25a), (4.25b), (4.59), and (4.57), it follows that

$$\begin{aligned}
C_\theta^T J C_\theta &= (\hat{C}_1 - D_{1\cdot}D_{2\cdot}^{-1}\hat{C}_2)^T J(\hat{C}_1 - D_{1\cdot}D_{2\cdot}^{-1}\hat{C}_2) \\
&= \hat{C}_1^T J\hat{C}_1 - \hat{C}_2^T \hat{D}_{2\cdot}^{-T}(D^T C + D_{2\cdot}^T J\hat{C}_2) \\
&\quad +(C^T D + \hat{C}_2^T J D_{2\cdot})\hat{D}_{2\cdot}^{-1}\hat{C}_2 + \hat{C}_2^T J'\hat{C}_2 \\
&= \hat{C}_1^T J\hat{C}_1 - \hat{C}_2^T J'\hat{C}_2 - \hat{C}_2^T D_{2\cdot}^{-T} D^T C - C^T D D_{2\cdot}^{-1}\hat{C}_2 \\
&= C^T C + \hat{C}_2^T D_{2\cdot}^{-T} B^T P + PBD_{2\cdot}^{-1}\hat{C}_2.
\end{aligned}$$

From this identity and the first relation of (4.57), we can easily show (4.56a) using (4.58). Finally, using (4.25b), the relation (4.56b) is shown as

$$\begin{aligned}
D_\theta^T J C_\theta + B_\theta^T P &= D_{2\cdot}^{-T} D_{1\cdot}^T J(\hat{C}_1 - D_{1\cdot}D_{2\cdot}^{-1}\hat{C}_2) + D_{2\cdot}^{-T} B^T P \\
&= D_{2\cdot}^{-T}(D^T C + D_{2\cdot}^T J'\hat{C}_2) - J'\hat{C}_2 + D_{2\cdot}^{-T} B^T P = 0.
\end{aligned}$$

If Θ is (J, J')-lossless, Σ is lossless due to Lemma 4.4. In that case, $P \geq 0$. This establishes the second assertion.

In view of (4.58), we can represent

$$A_\theta = A - LC, \; C_\theta = VC,$$
$$L = \begin{bmatrix} 0 & B_1 D_{21}^{-1} \end{bmatrix}, \; V = \begin{bmatrix} I & -D_{11}D_{21}^{-1} \\ 0 & -D_{21}^{-1} \end{bmatrix}.$$

Since V is nonsingular, (A_θ, C_θ) is observable iff (A, C) is observable. In that case, the observability gramian P in (4.57) is positive definite due to Lemma 2.14. Thus, we have established the final assertion. ∎

Example 4.4
A system

$$\Theta(s) = \begin{bmatrix} \dfrac{s-2}{s+2} & 0 \\ 0 & \dfrac{s+1}{s-1} \end{bmatrix}$$

is $J_{1,1}$-lossless. Indeed, for $J = \text{diag}[1, -1]$, we have

$$\Theta^\sim(s) J \Theta(s) = \begin{bmatrix} \dfrac{-s-2}{-s+2} & 0 \\ 0 & \dfrac{-s+1}{-s-1} \end{bmatrix} \begin{bmatrix} \dfrac{s-2}{s+2} & 0 \\ 0 & -\dfrac{s+1}{s-1} \end{bmatrix} = J,$$

$$\Theta^*(s)J\Theta(s) = J - \begin{bmatrix} \dfrac{8\sigma}{|s|^2+4\sigma+4} & 0 \\ 0 & \dfrac{4\sigma}{|s|^2-2\sigma+1} \end{bmatrix} \leq J,$$

$$\sigma = \text{Re } s > 0.$$

Now $\Theta(s)$ has a realization

$$\Theta(s) = \left[\begin{array}{cc|cc} -2 & 0 & -4 & 0 \\ 0 & 1 & 0 & 2 \\ \hline 1 & 0 & 1 & 0 \\ 0 & 1 & 0 & 1 \end{array}\right].$$

It is straightforward to see that

$$P = \begin{bmatrix} 1/4 & 0 \\ 0 & 1/2 \end{bmatrix} > 0$$

satisfies (4.56).

Example 4.5
A system

$$\Theta(s) = \begin{bmatrix} \dfrac{s+1}{s-1} & 0 \\ 0 & \dfrac{s-2}{s+2} \end{bmatrix}$$

is obtained by swapping its diagonal elements of the system in Example 4.4. From

$$J - \Theta^*(s)J\Theta(s) = -\begin{bmatrix} \dfrac{4\sigma}{|s|^2-2\sigma+1} & 0 \\ 0 & \dfrac{8\sigma}{|s|^2+4\sigma+4} \end{bmatrix} \leq 0,$$

we know that $\Theta(s)$ is *not* J-lossless.

Example 4.6
A slightly more complex example of a J-lossless system is given by

$$\Theta(s) = \frac{1}{s^2+s-2}\begin{bmatrix} s^2 - \dfrac{25}{8}s + 2 & \dfrac{7}{8}s \\ -\dfrac{7}{8}s & s^2 + \dfrac{25}{8}s + 2 \end{bmatrix}.$$

Its state-space realization is given by

$$\Theta(s) = \left[\begin{array}{cc|cc} 0 & -1 & -1 & -1 \\ -2 & -1 & -5/2 & 3/2 \\ \hline 1 & 5/4 & 1 & 0 \\ -1 & 3/4 & 0 & 1 \end{array}\right].$$

A direct computation proves that

$$P = \begin{bmatrix} 1 & 0 \\ 0 & 1/2 \end{bmatrix} > 0$$

satisfies (4.56) in this case.

Now we state several important properties of (J, J')-lossless matrices. A direct consequence of Theorem 4.5 is as follows.

LEMMA 4.6 *Any (J_{mr}, J_{pr})-lossless matrix Θ is always represented as*

$$\Theta = \Theta_0 D, \tag{4.61}$$

where Θ_0 is a J_{mr}-lossless matrix and D is a constant (J_{mr}, J_{pr})-unitary matrix.

Proof. Let (4.54) be an observable realization of Θ, and let $P > 0$ be a matrix satisfying (4.56). From (4.56b), $B_\theta = -P^{-1}C_\theta^T J_{mr} D_\theta$. Therefore, taking $D = D_\theta$ and

$$\Theta_0 := \left[\begin{array}{c|c} A_\theta & -P^{-1}C_\theta^T J_{mr} \\ \hline C_\theta & I \end{array}\right],$$

the representation (4.59) holds. ∎

Lemma 4.6 implies that any (J, J')-lossless system Θ is represented as

$$\Theta = \left[\begin{array}{c|c} A_\theta & -P^{-1}C_\theta^T J \\ \hline C_\theta & I \end{array}\right] D, \tag{4.62}$$

where P is a positive definite solution of

$$PA_\theta + A_\theta^T P + C_\theta^T J C_\theta = 0. \tag{4.63}$$

From this representation, we have the following result which demonstrates a salient characteristic feature of (J, J')-lossless systems.

LEMMA 4.7 *If λ is a zero of (J, J')-lossless system Θ, then, $-\lambda$ is a pole of Θ.*

Proof. Let Θ be given by (4.54) where P satisfies (4.56). If λ is a zero of Θ, there exist vectors ξ and η such that

$$(A_\theta - \lambda I)\xi - P^{-1}C_\theta^T J D_\theta \eta = 0, \;\; C_\theta \xi + D_\theta \eta = 0,$$

where we used (4.56b). Eliminating η yields

$$(A_\theta + P^{-1}C_\theta^T J C_\theta - \lambda I)\xi = 0,$$

which is identical to $(A_\theta^T + \lambda I)P\xi = 0$ due to (4.56a). Since $P > 0$, we conclude that $-\lambda$ is an eigenvalue of A_θ. Thus the assertion has been established. ∎

Given a (J_{mr}, J_{pr})-lossless system Θ with $m > p$, Θ_1 is said to be a *complement* of Θ if

$$\Theta_0 = \begin{bmatrix} \Theta_1 & \Theta \end{bmatrix}$$

is J_{mr}-lossless. Obviously, a complement Θ_1 of Θ is (J_{mr}, I_{m-p})-lossless and satisfies

$$\Theta^\sim J \Theta_1 = 0.$$

For a constant (J_{mr}, J_{pr})-unitary matrix, the complement always exists. Indeed, let D be a (J_{mr}, J_{pr})-unitary, that is,

$$D^T J_{mr} D = J_{pr}, \tag{4.64}$$

and write $U := J_{mr} - DJ_{pr}D^T$. Due to (4.64), $U^T J_{mr} U = U^T = U$, and $D^T J_{mr} U = 0$. Therefore, we have

$$\begin{bmatrix} U^T \\ D^T \end{bmatrix} J_{mr} \begin{bmatrix} U & D \end{bmatrix} = \begin{bmatrix} U & 0 \\ 0 & J_{pr} \end{bmatrix}.$$

Since $\begin{bmatrix} U & D \end{bmatrix}$ is nonsingular, the inertia theorem implies that $U \geq 0$ and rank$U = m - p$. Therefore, we can write $U = D_1 D_1^T$ for some D_1 with full column rank $m - p$. It is obvious that $D_1^T J_{mr} D_1 = I_{m-p}$ and $D_1^T J_{mr} D = 0$. Therefore, D_1 is a complement of D.

The following result is a direct consequence of the preceding reasoning.

LEMMA 4.8 *Any (J_{mr}, J_{pr})-lossless matrix has its complement.*

Proof. Let Θ be a (J_{mr}, J_{pr})-lossless matrix. Then Lemma 4.6 allows a representation (4.61). Let D_1 be a complement of D. It is obvious that $\Theta_1 = \Theta_0 D_1$ is a complement of Θ. ∎

It is well known that a J-lossless matrix can be factorized according to the decomposition of its set of poles. The following result extends this fact to general (J, J')-lossless matrices.

LEMMA 4.9 *Let $\mathbf{C} = \mathbf{C}_1 \cup \mathbf{C}_2$ be a division of the complex plane into two disjoint sets $\mathbf{C}_1$ and $\mathbf{C}_2$. Any (J_{mr}, J_{pr})-lossless matrix $\Theta(s)$ is represented as a product*

$$\Theta(s) = \Theta_1(s)\Theta_2(s), \tag{4.65}$$

where $\Theta_1(s)$ is a J_{mr}-lossless matrix whose poles are in $\mathbf{C}_1$ and $\Theta_2(s)$ is a (J_{mr}, J_{pr})-lossless matrix whose poles are in $\mathbf{C}_2$.

Proof. We can always find a minimal state-space realization of $\Theta(s)$ in the form

$$\Theta(s) = \left[\begin{array}{cc|c} A_1 & 0 & B_1 \\ 0 & A_2 & B_2 \\ \hline C_1 & C_2 & D \end{array}\right], \tag{4.66}$$

where $\sigma(A_1) \subset \mathbf{C}_1$ and $\sigma(A_2) \subset \mathbf{C}_2$. Due to Theorem 4.5, there exists a solution

$$P = \begin{bmatrix} P_{11} & P_{12} \\ P_{12}^T & P_{22} \end{bmatrix} > 0$$

of the equations

$$P\begin{bmatrix} A_1 & 0 \\ 0 & A_2 \end{bmatrix} + \begin{bmatrix} A_1^T & 0 \\ 0 & A_2^T \end{bmatrix} P + \begin{bmatrix} C_1^T \\ C_2^T \end{bmatrix} J \begin{bmatrix} C_1 & C_2 \end{bmatrix} = 0, \tag{4.67}$$

$$P\begin{bmatrix} B_1 \\ B_2 \end{bmatrix} + \begin{bmatrix} C_1^T \\ C_2^T \end{bmatrix} JD = 0. \tag{4.68}$$

We can show that the two systems given by

$$\Theta_1(s) := \left[\begin{array}{c|c} A_1 & -P_{11}^{-1}C_1^T J \\ \hline C_1 & I \end{array}\right], \tag{4.69}$$

$$\Theta_2(s) := \left[\begin{array}{c|c} A_2 & B_2 \\ \hline C_2 - C_1 P_{11}^{-1} P_{12} & D \end{array}\right] \tag{4.70}$$

satisfy (4.65). Indeed, from the concatenation rule (2.13), it follows that

$$\Theta_1(s)\Theta_2(s) = \left[\begin{array}{cc|c} A_1 & P_{11}^{-1}C_1^T J(C_1 P_{11}^{-1} P_{12} - C_2) & -P_{11}^{-1}C_1^T JD \\ 0 & A_2 & B_2 \\ \hline C_1 & C_2 - C_1 P_{11}^{-1} P_{12} & D \end{array}\right].$$

Taking the similarity transformation with the transformation matrix

$$T = \begin{bmatrix} I & -P_{11}^{-1}P_{12} \\ 0 & I \end{bmatrix}, \quad T^{-1} = \begin{bmatrix} I & P_{11}^{-1}P_{12} \\ 0 & I \end{bmatrix}$$

and using the relations (4.67) and (4.68) verify that the product $\Theta_1(s)\Theta_2(s)$ really coincides with $\Theta(s)$ as given by (4.66). It is obvious that $\Theta_1(s)$ is J-lossless, because P_{11} satisfies Equations (4.56) in Theorem 4.5 for $\Theta_1(s)$. As for $\Theta_2(s)$, it is not difficult to see that the Schur complement $P_{22} - P_{12}^T P_{11}^{-1} P_{12}$ of P satisfies Equations (4.56) in Theorem 4.5 for the realization (4.70). ∎

4.5 Dual (J, J')-Lossless Systems

In the preceding section, we showed that the *chain-scattering representation* of a lossless system is (J, J')-lossless. In this section, we investigate the corresponding property of *dual chain-scattering representation.* As we see later, the dual chain-scattering representation $\Psi = DCHAIN\,(\Sigma)$ of a lossless system Σ satisfies

$$\Psi(s) J_{mr} \Psi^{\sim}(s) = J_{mq} \quad \forall s \tag{4.71}$$

$$\Psi(s) J_{mr} \Psi^{*}(s) \geq J_{mq}, \quad \mathrm{Re}\, s \geq 0 \tag{4.72}$$

where $m = \dim(a_1) = \dim(b_2), q = \dim(a_2), r = \dim(b_1)$ in the representation (4.1) of Σ. The assumption $\dim(a_1) = \dim(b_2)$ is due to the invertibility of Σ_{12}. A matrix Ψ satisfying (4.71) and (4.72) is called *dual* (J_{mr}, J_{mq})*-lossless* or simply, *dual* (J, J')*-lossless.* It is clear that if Θ is J-lossless, then $\Psi = \Theta^{\sim}$ is dual J-lossless.

LEMMA 4.10 *A matrix $\Psi(s)$ is dual (J_{mr}, J_{mq})-lossless, if and only if it is the dual chain-scattering representation of a dual lossless system.*

Proof. Let

$$\Psi = \begin{bmatrix} \Psi_{11} & \Psi_{12} \\ \Psi_{21} & \Psi_{22} \end{bmatrix}.$$

Due to (4.71), we have

$$\Psi_{11}\Psi_{11}^* - \Psi_{12}\Psi_{12}^* \geq I_m.$$

This implies that Ψ_{11} is invertible. Let

$$N = \begin{bmatrix} -\Psi_{12} & I \\ \Psi_{22} & 0 \end{bmatrix}, \quad M = \begin{bmatrix} \Psi_{11} & 0 \\ -\Psi_{21} & I \end{bmatrix}$$

and define $\Sigma := M^{-1}N$. Then, due to (4.35), $\Psi = DCHAIN\,(\Sigma)$. It is easy to prove the identity

$$MM^* - NN^* = J_{mq}\Psi J_{mr}\Psi^* J_{mq} - J_{mq},$$

from which the dual losslessness of Σ follows immediately. The converse can be proven analogously. ∎

The state-space characterization of *dual* (J, J')*-lossless* systems can be obtained in an analogous way as in the previous section. Let

$$\Psi(s) = \left[\begin{array}{c|c} A_\psi & B_\psi \\ \hline C_\psi & D_\psi \end{array}\right] \in \mathbf{RL}^\infty_{(m+q)\times(m+r)} \tag{4.73}$$

be a state-space realization.

LEMMA 4.11 *A matrix $\Psi(s)$ given by (4.73) is dual (J_{mr}, J_{mq})-unitary if and only if*

$$D_\psi J_{mr} D_\psi^T = J_{mq} \tag{4.74}$$

and there exists a matrix Q such that

$$QA_\psi^T + A_\psi Q - B_\psi^T J_{mr} B_\psi = 0, \tag{4.75a}$$

$$D_\psi J B_\psi^T - C_\psi Q = 0. \tag{4.75b}$$

The solution Q satisfies $Q \geq 0$ iff Ψ is dual (J_{mr}, J_{mq})-lossless. The solution satisfies $Q > 0$ iff (A_ψ, B_ψ) is controllable.

Proof. Assume that Ψ is dual (J_{mr}, J_{mq})-lossless. Due to Lemma 4.10, there exists a dual unitary system Σ such that $\Psi = DCHAIN\,(\Sigma)$. Let

$$\Sigma = \left[\begin{array}{c|c} A & B \\ \hline C & D \end{array}\right]$$

be a realization of Σ. Since Σ is dual lossless, Theorem 3.14 implies that $DD^T = I$ and there exists a matrix Q such that

$$QA^T + AQ + BB^T = 0,\ QC^T + BD^T = 0 \tag{4.76}$$

Due to Lemma 4.3,

$$A_\psi = A - \hat{B}_2 D_{\cdot 2}^{-1} C,\ B_\psi = \hat{B}_1 - \hat{B}_2 D_{\cdot 2}^{-1} D_{\cdot 1},$$
$$C_\psi = D_{\cdot 2}^{-1} C,\ D_\psi = D_{\cdot 2}^{-1} D_{\cdot 1},$$

where $\hat{B}_1, \hat{B}_2, D_{\cdot 1}, D_{\cdot 2}$ are given by (4.41). Write

$$J = J_{mr},\ J' = J_{mq},$$

for simplicity. Since $DD^T = I$, we have, from (4.46),

$$D_{\cdot 2} J' D_{\cdot 2}^T = D_{\cdot 1} J D_{\cdot 1}^T,$$

which implies (4.74). Due to (4.44), (4.45), and (4.76), we have

$$\begin{aligned} B_\psi J B_\psi^T &= (\hat{B}_1 - \hat{B}_2 D_{\cdot 2}^{-1} D_{\cdot 1}) J (\hat{B}_1 - \hat{B}_2 D_{\cdot 2}^{-1} D_{\cdot 1})^T \\ &= \hat{B}_1 J \hat{B}_1 - \hat{B}_2 J' \hat{B}_2^T + \hat{B}_2 D_{\cdot 2}^{-1} D B^T + B D^T D_{\cdot 2}^{-T} \hat{B}_2^T \\ &= -BB^T - \hat{B}_2 D_{\cdot 2}^{-1} C Q - Q C^T D_{\cdot 2}^{-T} \hat{B}_2^T \\ &= A_\psi Q + Q A_\psi^T, \end{aligned}$$

which establishes (4.75a). From (4.76) and (4.43), it follows that

$$\begin{aligned} D_\psi J B_\psi^T - C_\psi Q &= D_{\cdot 2}^{-1} D_{\cdot 1} J (\hat{B}_1^T - D_{\cdot 1}^T D_{\cdot 2}^{-T} \hat{B}_2^T) - D_{\cdot 2}^{-1} C Q \\ &= D_{\cdot 2}^{-1} (D_{\cdot 2} J' \hat{B}_2^T - D B^T) - J' \hat{B}_2^T + D_{\cdot 2}^{-1} D B^T = 0, \end{aligned}$$

which establishes (4.75b).

If Ψ is dual (J, J')-lossless, then Σ is dual lossless. In that case, Q in (4.76) satisfies $Q \geq 0$, which establishes the second assertion. As in the proof of Theorem 4.5, we can show that (A_ψ, B_ψ) is controllable iff (A, B) is controllable. The last assertion follows immediately from Lemma 3.2. ∎

Comparison of (4.56) and (4.75) yields the following result.

LEMMA 4.12 *A matrix Θ is J-lossless iff $\Psi = \Theta^\sim$ is dual J-lossless.*

4.6 Feedback and Terminations

Consider again a system Σ described in (4.1). If a_2 is fed back to b_2 by

$$b_2 = Sa_2, \tag{4.77}$$

then the transfer function from b_1 to a_1 is given by

$$a_1 = \Phi b_1 \tag{4.78}$$

$$\Phi = \Sigma_{11} + \Sigma_{12}S(I - \Sigma_{22}S)^{-1}\Sigma_{21}. \tag{4.79}$$

The relation (4.79) is sometimes called a *linear fractional transformation* in the control literature and is denoted as

$$LF\ (\Sigma; S) := \Sigma_{11} + \Sigma_{12}S(I - \Sigma_{22}S)^{-1}\Sigma_{21}. \tag{4.80}$$

The same relation can be described in terms of $\Theta = CHAIN\ (\Sigma)$ defined as

$$\begin{bmatrix} a_1 \\ b_1 \end{bmatrix} = \Theta \begin{bmatrix} b_2 \\ a_2 \end{bmatrix} = \begin{bmatrix} \Theta_{11} & \Theta_{12} \\ \Theta_{21} & \Theta_{22} \end{bmatrix} \begin{bmatrix} b_2 \\ a_2 \end{bmatrix}. \tag{4.81}$$

Substitution of (4.77) in (4.81) yields

$$\begin{bmatrix} a_1 \\ b_1 \end{bmatrix} = \begin{bmatrix} \Theta_{11}S + \Theta_{12} \\ \Theta_{21}S + \Theta_{22} \end{bmatrix} a_2.$$

Therefore, Φ in (4.78) is given by

$$\Phi = (\Theta_{11}S + \Theta_{12})(\Theta_{21}S + \Theta_{22})^{-1}.$$

This relation is denoted as

$$HM\ (\Theta; S) := (\Theta_{11}S + \Theta_{12})(\Theta_{21}S + \Theta_{22})^{-1}. \tag{4.82}$$

The symbol *HM* stands for "*Homographic Transformation*," which was used in classical circuit theory. The representation (4.82) is also called a linear fractional transformation, but in this book we adopt the words *Homographic Transformation* according to [15], in order to distinguish it from (4.80). The Homographic Transformation represents a feedback in terms of chain-scattering representations. In the context of circuit theory, it represents the "termination" of a port by a load. Figure 4.9

represents *LF* and *HM*. The termination of a chain-scattering representation is equivalent to the feedback of an input/output system.

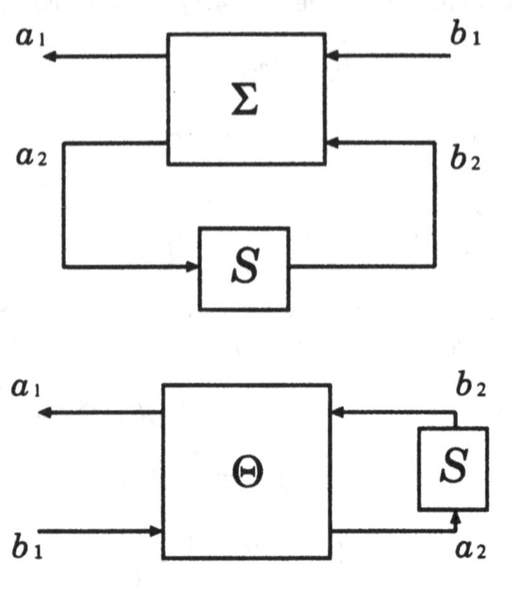

Figure 4.9 *LF* and *HM*.

Now we list some properties of *HM*.

LEMMA 4.13 *$HM\ (\Theta; S)$ satisfies the following properties.*

(i) $HM\ (CHAIN(\Sigma)\ ;\ S) = LF\ (\Sigma; S)$.

(ii) $HM\ (I; S) = S$.

(iii) $HM\ (\Theta_1; HM\ (\Theta_2; S)) = HM\ (\Theta_1\Theta_2; S)$.

(iv) If $HM\ (\Theta; S) = F$ *and* Θ^{-1} *exists,* $S = HM\ (\Theta^{-1}; F)$.

Proof. (i) is a defining property of *HM*. (ii) is obvious from the definition (4.82). Statement (iii) is clear from the concatenation property of chain-scattering representation described in Figure 4.10. Property (iv) is a direct consequence of (ii) and (iii). ∎

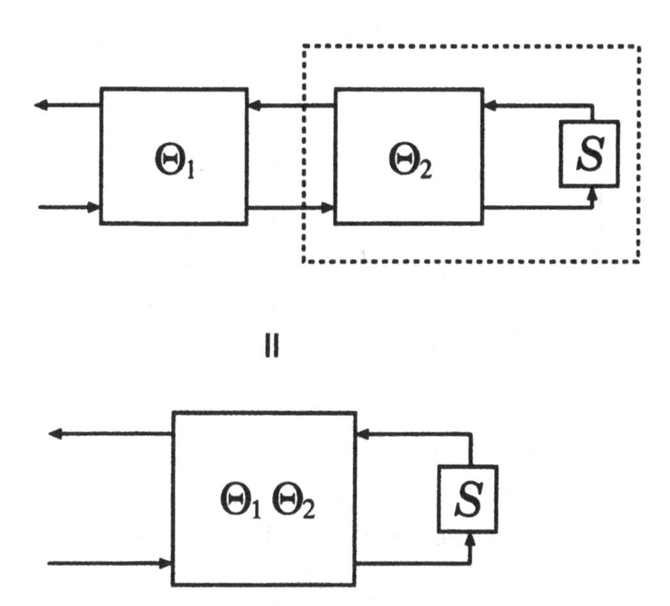

Figure 4.10 Cascade Structure of *HM*.

Now we show a few examples of *HM*.

Example 4.7 Parameterization of Stabilizing Controllers
Assume that the system

$$\Sigma : \begin{array}{l} \dot{x} = Ax + Bu \\ y = Cx \end{array}$$

is stabilizable and detectable. Let F and H be gains of state feedback and output insertion such that $A + BF$ and $A + HC$ are both stable. Define

$$U = \left[\begin{array}{c|cc} A+BF & B & -H \\ \hline F & I & 0 \\ C & 0 & I \end{array}\right].$$

Then the controller

$$u = Ky$$

stabilizes the plant Σ iff it is represented by

$$K = HM\,(U;Q)$$

for some stable Q.

Example 4.8 Input Impedance of Electrical Circuit (Figure 4.11)

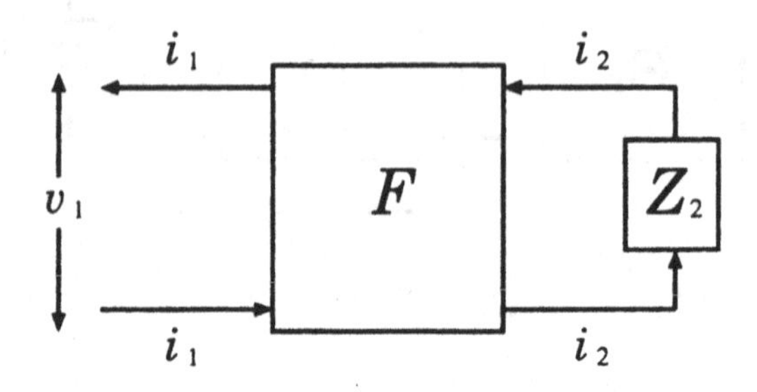

Figure 4.11 Electrical Circuit.

The input impedance of an electrical circuit described by a cascade matrix F terminated by an impedance Z_2 is given by

$$Z = HM\,(F; Z_2).$$

Example 4.9 Perturbed Linear Equation
Consider a linear equation

$$Ax = b.$$

If the (i,j)th element a_{ij} of A is perturbed by Δ, that is,

$$a_{ij} \to a_{ij} + \Delta,$$

the solution to the perturbed equation is represented as

$$x_k = HM\,(\begin{bmatrix} |P_{ij}| & |P| \\ |A_{ij}| & |A| \end{bmatrix}, \Delta),$$

where $|A_{ij}|$ is the (i,j)th cofactor of A and P is the matrix A with its kth column replaced by b.

Now we derive a state-space form of $HM\,(\Theta; S)$ in terms of state-space forms of Θ and S given, respectively, by

$$\Theta = \left[\begin{array}{c|cc} A & B_1 & B_2 \\ \hline C_1 & D_{11} & D_{12} \\ C_2 & D_{21} & D_{22} \end{array}\right], \quad S = \left[\begin{array}{c|c} A_s & B_s \\ \hline C_s & D_s \end{array}\right]. \tag{4.83}$$

The relation (4.3) is then described as

$$\begin{aligned}\dot{x} &= Ax + B_1 b_2 + B_2 a_2,\\ a_1 &= C_1 x + D_{11} b_2 + D_{12} a_2,\\ b_1 &= C_2 x + D_{21} b_2 + D_{22} a_2,\end{aligned}$$

and the relation (4.77) is described as

$$\begin{aligned}\dot{\xi} &= A_s \xi + B_s a_2,\\ b_2 &= C_s \xi + D_s a_2,\end{aligned}$$

where x and ξ are the states of Θ and S, respectively. Elimination of b_2 from the preceding relations yields

$$\begin{bmatrix} \dot{x} \\ \dot{\xi} \end{bmatrix} = \begin{bmatrix} A & B_1 C_s \\ 0 & A_s \end{bmatrix} \begin{bmatrix} x \\ \xi \end{bmatrix} + \begin{bmatrix} \hat{B} \\ B_s \end{bmatrix} a_2,$$

$$\begin{bmatrix} a_1 \\ b_1 \end{bmatrix} = \begin{bmatrix} C_1 & D_{11} C_s \\ C_2 & D_{21} C_s \end{bmatrix} \begin{bmatrix} x \\ \xi \end{bmatrix} + \begin{bmatrix} D_1 \\ D_2 \end{bmatrix} a_2,$$

where

$$\begin{bmatrix} \hat{B} \\ D_1 \\ D_2 \end{bmatrix} = \begin{bmatrix} B_1 & B_2 \\ D_{11} & D_{12} \\ D_{21} & D_{22} \end{bmatrix} \begin{bmatrix} D_s \\ I \end{bmatrix}. \tag{4.84}$$

From these relations, we obtain a state-space realization of $HM\,(\Theta; S)$ which represents the transfer function from b_1 to a_1 as

$$HM\,(\Theta; S) = \left[\begin{array}{c|c} A_c & B_c \\ \hline C_c & D_c \end{array}\right], \tag{4.85}$$

$$A_c = \begin{bmatrix} A & B_1 C_s \\ 0 & A_s \end{bmatrix} - \begin{bmatrix} \hat{B} \\ B_s \end{bmatrix} D_2^{-1} \begin{bmatrix} C_2 & D_{21} C_s \end{bmatrix}, \tag{4.86}$$

$$B_c = \begin{bmatrix} \hat{B} \\ B_s \end{bmatrix} D_2^{-1}, \quad C_c = \begin{bmatrix} C_1 - D_c C_2 & (D_{11} - D_c D_{21}) C_s \end{bmatrix},$$

$$D_c = D_1 D_2^{-1}, \tag{4.87}$$

subject to the condition that $D_2 = D_{21} D_s + D_{22}$ is invertible. This is the condition for the well-posedness of the feedback scheme of Figure 4.9.

Now, we define the internal stability of the terminated system of Figure 4.9

DEFINITION 4.14 *A terminated system of Figure 4.9 is said to be internally stable if the matrix A_c given in (4.86) is stable, where minimal realizations of Θ and S are given by (4.83).*

We prove a fundamental result that is strongly relevant to the H^∞ control problem.

THEOREM 4.15 *Assume that Θ is (J, J')-unitary. There exists a termination S such that $HM(\Theta; S) \in \mathbf{BH}^\infty$ iff Θ is (J, J')-lossless. In that case, $HM(\Theta; S) \in \mathbf{BH}^\infty$ iff $S \in \mathbf{BH}^\infty$.*

Proof. From the definition of $HM(\Theta; S)$, we have

$$HM(\Theta; S) = NM^{-1}$$
$$\begin{bmatrix} N \\ M \end{bmatrix} = \Theta \begin{bmatrix} S \\ I \end{bmatrix}.$$

Since Θ is (J, J')-unitary, we have

$$N^{\sim}(j\omega)N(j\omega) - M^{\sim}(j\omega)M(j\omega) = S^{\sim}(j\omega)S(j\omega) - I.$$

If $HM(\Theta; S) \in \mathbf{BH}^\infty$, $N^{\sim}(j\omega)N(j\omega) - M^{\sim}(j\omega)M(j\omega) < 0$ for each ω. Therefore,

$$S^{\sim}(j\omega)S(j\omega) < I, \quad \forall\omega.$$

Assume that S has a state-space realization (4.83). Due to Theorem 3.9, there exists a stabilizing solution X_s of ARE,

$$\begin{aligned} X_sA_s + A_s^TX_s + (C_s^TD_s + X_sB_s)R^{-1}(D_s^TC_s + B_s^TX_s) + C_s^TC_s = 0 \\ R := I - D_s^TD_s > 0. \end{aligned} \tag{4.88}$$

Assume that Θ has a state-space realization (4.83). Since Θ is (J, J')-unitary, Theorem 4.5 implies that there exists a solution P to the equations

$$PA + A^TP + C_1^TC_1 - C_2^TC_2 = 0,$$
$$\begin{bmatrix} D_{11}^T & -D_{21}^T \\ D_{12}^T & -D_{22}^T \end{bmatrix} \begin{bmatrix} C_1 & D_{11} & D_{12} \\ C_2 & D_{21} & D_{22} \end{bmatrix} = \begin{bmatrix} B_1^TP & I & 0 \\ B_2^TP & 0 & -I \end{bmatrix}.$$

These relations yield

$$
\begin{aligned}
D_2^T D_2 - D_1^T D_1 &= I - D_s^T D_s, \\
D_{21}^T D_2 + D_s &= D_{11}^T D_1, \\
P\hat{B} &= C_2^T D_2 - C_1^T D_1,
\end{aligned}
$$

where $\hat{B}$, D_1, and D_2 are given by (4.84). From these relations, we have

$$
D_{11}^T D_c = D_{21}^T + D_s D_2^{-1}, \; D_2^T (I - D_c^T D_c) D_2 = I - D_s^T D_s,
$$

$$
\begin{aligned}
&\begin{bmatrix} C_2^T \\ C_s^T D_{21}^T \end{bmatrix} (I - D_c^T D_c) \begin{bmatrix} C_2 & D_{21} C_s \end{bmatrix} + C_c^T C_c \\
&\quad = \begin{bmatrix} -P\hat{B} D_2^{-1} C_2 - C_2^T D_2^{-T} \hat{B}^T P & C_1^T (D_{11} - D_c D_{21}) C_s \\ C_s^T (D_{11} - D_c D_{21})^T C_1 & D_s D_2^{-1} D_{21} - D_{21}^T D_2^{-T} D_s \end{bmatrix} \\
&\quad + \begin{bmatrix} C_1^T C_1 - C_2^T C_2 & C_2^T D_2^{-T} D_s C_s \\ -C_s^T D_s^T D_2^{-1} C_2 & I \end{bmatrix},
\end{aligned}
$$

using the expressions (4.87) and (4.84). It follows that

$$
\begin{aligned}
&C_c^T D_c + \begin{bmatrix} P & 0 \\ 0 & X_s \end{bmatrix} B_c = \\
&\quad + \begin{bmatrix} C_2^T \\ C_s^T D_{21}^T \end{bmatrix} (I - D_c^T D_c) + \begin{bmatrix} 0 \\ X_s B_s + C_s^T D_s \end{bmatrix} D_2^{-1}.
\end{aligned}
$$

Simple but lengthy algebraic manipulations based on the preceding relations yield

$$
\begin{aligned}
&\begin{bmatrix} P & 0 \\ 0 & X_s \end{bmatrix} A_c + A_c^T \begin{bmatrix} P & 0 \\ 0 & X_s \end{bmatrix} \\
&+ \left(C_c^T D_c + \begin{bmatrix} P & 0 \\ 0 & X_s \end{bmatrix} B_c \right) (I - D_c^T D_c)^{-1} \left(D_c^T C_c + B_c^T \begin{bmatrix} P & 0 \\ 0 & X_s \end{bmatrix} \right) \\
&+ C_c^T C_c = 0.
\end{aligned} \tag{4.89}
$$

Stability of A_c implies $P \geq 0$, $X_s \geq 0$, which is equivalent to $HM\,(\Theta; S) \in \mathbf{BH}^\infty$ according to Theorem 3.9. ∎

It is worth noting that the matrix A_c given in (4.86) is the A-matrix of the inverse of

$$\Theta_{21}S + \Theta_{22} = \begin{bmatrix} \Theta_{21} & \Theta_{22} \end{bmatrix} \begin{bmatrix} S \\ I \end{bmatrix}$$

$$= \left[\begin{array}{c|cc} A & B_1 & B_2 \\ \hline C_2 & D_{21} & D_{22} \end{array}\right] \cdot \left[\begin{array}{c|c} A_S & B_S \\ \hline C_S & D_S \\ 0 & I \end{array}\right]$$

$$= \left[\begin{array}{cc|c} A & B_1C_S & B_1D_S + B_2 \\ 0 & A_S & B_S \\ \hline C_2 & D_{21}C_S & D_{21}D_S + D_{22} \end{array}\right].$$

Based on this observation, we can show the following result which is intuitively clear.

LEMMA 4.16 *Assume that a terminated system $HM\,(\Theta; S)$ is internally stable. Then, for any unimodular Π, $HM\,(\Theta\Pi; HM\,(\Pi^{-1}; S))$ is internally stable.*

Proof. Since $HM\,(\Theta\Pi; HM\,(\Pi^{-1}; S)) = HM\,(\Theta; S)$, the A-matrix of the state equation of Figure 4.12(a) is composed of the A-matrix of the system equation describing the system of Figure 4.12(b) and those of Π and Π^{-1}. Since Figure 4.12(b) is internally stable from the assumption and Π is unimodular, both parts of the A-matrix of the state equation of $HM\,(\Theta\Pi; HM\,(\Pi^{-1}; S))$ are stable. ∎

Now we introduce a dual notion of *HM*, which corresponds to $\Psi = DCHAIN\,(\Sigma)$. The relation (4.27) is written as

$$\begin{bmatrix} b_2 \\ a_2 \end{bmatrix} = \Psi \begin{bmatrix} a_1 \\ b_1 \end{bmatrix} = \begin{bmatrix} \Psi_{11} & \Psi_{12} \\ \Psi_{21} & \Psi_{22} \end{bmatrix} \begin{bmatrix} a_1 \\ b_1 \end{bmatrix}. \tag{4.90}$$

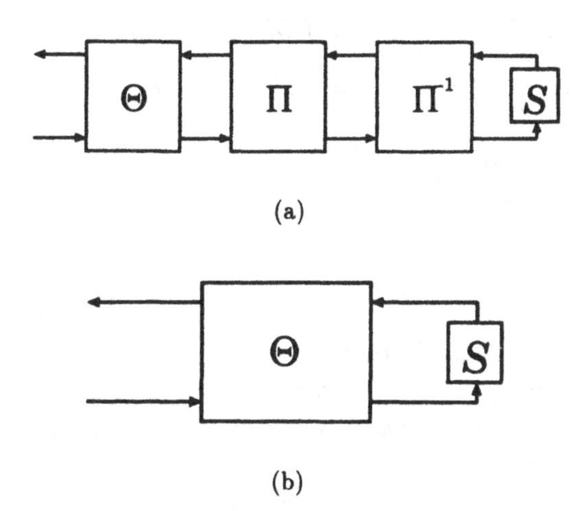

Figure 4.12 Equivalent Block Diagrams.

The feedback (4.77) is written as

$$\begin{bmatrix} I & -S \end{bmatrix} \begin{bmatrix} b_2 \\ a_2 \end{bmatrix} = 0.$$

Therefore, due to (4.90), we have

$$(\Psi_{11} - S\Psi_{21})a_1 + (\Psi_{12} - S\Psi_{22})b_1 = 0.$$

It follows that

$$a_1 = -(\Psi_{11} - S\Psi_{21})^{-1}(\Psi_{12} - S\Psi_{22})b_1.$$

This is an alternative representation of Φ in (4.79), and is denoted by

$$DHM\ (\Psi; S) := -(\Psi_{11} - S\Psi_{21})^{-1}(\Psi_{12} - S\Psi_{22}), \qquad (4.91)$$

which is obviously a dual of *HM*. The following lemma lists some properties of *DHM*.

LEMMA 4.17 *DHM* $(\Psi; S)$ *satisfies the following properties.*

(i) $DHM\ (DCHAIN\ (\Sigma); S) = LF\ (\Sigma; S)$.

(ii) $DHM\ (I; S) = S$.

(iii) $DHM(\Psi_1; DHM(\Psi_2; S)) = DHM(\Psi_2\Psi_1; S)$.

(iv) If $DHM(\Psi; S) = G$ *and* Ψ^{-1} *exists,* $S = DHM(\Psi^{-1}; G)$.

(v) If a left inverse $\Psi^\dagger$ *of* Ψ *exists, then* $DHM(\Psi; S) = HM(\Psi^\dagger; S)$.

Notes

The main subjects of this chapter were the chain-scattering representation and the J-lossless system. It is not clear when the chain-scattering representation was initially used, but certainly it has been used extensively in various fields of engineering as a method of representing the relationship between the power ports, as was shown in Example 4.2, and/or the scattering properties of physical systems. Its advantage over the usual input/output representation is so obvious at least in dealing with feedback connection that it is strange that it has not been popular in the control community so far. The use of the chain-scattering representation in this book was influenced by the behavioral approach to dynamical systems proposed by Willems [99][100] which made us aware of the alternatives to represent dynamical systems beyond the input/output representation. The papers by Green [35] and by Tsai and Tsai [96] used almost the same framework as the chain-scattering representation. However, they avoided this representation because it requires the invertibility of P_{21} which is not generally the case. We circumvent this difficulty by introducing the augmentation of plants which is another salient feature of the approach in this book. A concise treatment of the chain-scattering representation is found in [61].

The dual chain-scattering representation introduced in Section 4.3 is essentially new. The equivalence of the inverse and the dual is a new observation made in this book. The reader can see the most fundamental representation of the duality in linear system theory. The notion of chain-scattering representation actually generates operator theory in Krein space which is a Hilbert space with indefinite metric [6][1]. It requires the establishment of a system theory in the framework of Krein space [5]. The notice of J-losslessness is actually the representation of losslessness in the framework of Krein space. Extensive analysis of the J-lossless system is found in [17][18]. The factorization of J-lossless systems stated in Lemma 4.9 was initially found by Potapov [79]. It is important in reflecting the cascade structure of chain-scattering representations. The notation $HM(\Theta; S)$ introduced here was used in [57]. The properties of this transformation in the case of constant matrices were extensively investigated by Siegel [90].

Recently, an extension of the chain-scattering representation to general plants with noninvertible D_{21} was obtained in [102][103], where the plant (4.18) was written by a descriptor form

$$\begin{bmatrix} I & 0 \\ 0 & 0 \end{bmatrix} \begin{bmatrix} \dot{x} \\ \dot{b}_1 \end{bmatrix} = \begin{bmatrix} A & B_1 \\ C_2 & D_{21} \end{bmatrix} \begin{bmatrix} x \\ b_1 \end{bmatrix} + \begin{bmatrix} B_2 & 0 \\ 0 & -I \end{bmatrix} \begin{bmatrix} b_2 \\ a_2 \end{bmatrix}, \tag{4.92a}$$

$$\begin{bmatrix} a_1 \\ b_1 \end{bmatrix} = \begin{bmatrix} C_1 & D_{11} \\ 0 & I \end{bmatrix} \begin{bmatrix} x \\ b_1 \end{bmatrix} + \begin{bmatrix} D_{12} & 0 \\ 0 & 0 \end{bmatrix} \begin{bmatrix} b_2 \\ a_2 \end{bmatrix}. \tag{4.92b}$$

This is actually a chain-scattering representation (4.3). Almost all the results obtained in this chapter can be extended to the descriptor form representation (4.92).

Problems

[1] For the matrices introduced in (4.22) and (4.41), prove the identities:

$$D_{\cdot 1}D_{1\cdot} = D_{\cdot 2}D_{2\cdot} = \begin{bmatrix} 0 & -D_{12} \\ D_{21} & 0 \end{bmatrix}.$$
$$B = \hat{B}_1 D_{1\cdot} - \hat{B}_2 D_{2\cdot}.$$
$$C = D_{\cdot 2}\hat{C}_2 - D_{\cdot 1}\hat{C}_1.$$
$$\hat{B}_1\hat{C}_1 = \hat{B}_2\hat{C}_2 = 0.$$

[2] Assume that $a, b, c, d \in \mathbf{R}$ satisfies

$$\det \begin{bmatrix} a & b \\ c & d \end{bmatrix} = ad - bc = 1.$$

Prove that

$$g = HM\left(\begin{bmatrix} a & b \\ c & d \end{bmatrix}, h\right)$$

maps $H = \left\{h \in \mathbf{C};\ \mathrm{Im}(h) = \frac{1}{2j}(h - \bar{h}) > 0\right\}$ onto itself.

[3] Let

$$U := \begin{bmatrix} 0 & I_m \\ -I_m & 0 \end{bmatrix}$$

and write

$$\mathcal{H} = \left\{ H \in \mathbf{C}^{m\times m};\ \frac{1}{2j}(H - \bar{H}^T) > 0 \right\}.$$

Prove that $G = HM\,(T; H)$ maps $\mathcal{H}$ onto itself if $T \in \mathbf{C}^{2m\times 2m}$ satisfies

$$\bar{T}^T U T = U.$$

[4] Assume that G has a left inverse $G^\dagger$; that is, $G^\dagger G = I$. Show that

$$HM\,(G; S) = DHM\,(G^\dagger; S)$$

for each S.

[5] Assume that the signal a_2 in Figure 4.2 is divided into two components a_2' and a_2'' as shown in Figure 4.13, and G is written conformably as

$$G = \begin{bmatrix} G_{11} & G_{12} & G_{13} \\ G_{21} & G_{22} & G_{23} \end{bmatrix}.$$

Show that

$$HM\left(G; \begin{bmatrix} K_1 & K_2 \end{bmatrix}\right) = \begin{bmatrix} G_{11}K_1 + G_{12} & G_{11}K_2 + G_{13} \end{bmatrix} \cdot \begin{bmatrix} G_{21}K_1 + G_{22} & G_{21}K_2 + G_{23} \end{bmatrix}^{-1}.$$

Show also that if $K_2 = 0$,

$$HM\left(G\begin{bmatrix} I & 0 \\ Q & R \end{bmatrix}; \begin{bmatrix} K_1 & 0 \end{bmatrix}\right) = HM\left(G; \begin{bmatrix} K_1 & 0 \end{bmatrix}\right)$$

for any Q and R with R being nonsingular.

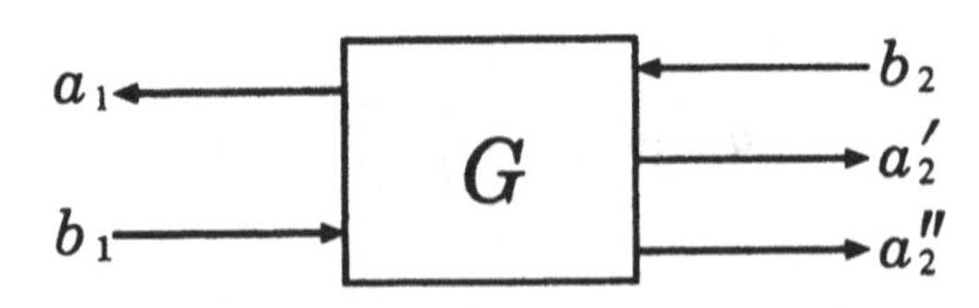

Figure 4.13.

[**6**] Obtain the dual result of Problem [5].

[**7**] Assume that

$$CHAIN(P) = \left[\begin{array}{c|cc} A & B_1 & B_2 \\ \hline C_1 & D_{11} & D_{12} \\ C_2 & D_{21} & D_{22} \end{array}\right].$$

Show that

$$P = \left[\begin{array}{c|cc} A - B_2 D_{22}^{-1} C_2 & B_2 D_{22}^{-1} & B_1 - B_2 D_{22}^{-1} D_{21} \\ \hline C_1 - D_{12} D_{22}^{-1} C_2 & D_{12} D_{22}^{-1} & D_{11} - D_{12} D_{22}^{-1} D_{21} \\ -D_{22}^{-1} C_2 & D_{22}^{-1} & -D_{22}^{-1} D_{21} \end{array}\right].$$

[**8**] Prove that if $\Pi(s)$ is J-lossless and unimodular, it is a constant J-unitary matrix.

[**9**] Let L be a constant $(m+r)\times(p+q)$ matrix satisfying $L^T J_{mr} L = J_{pq}$. Show that

$$J_{mr} - L J_{pq} L^T = U^T \begin{bmatrix} I_{m-p} & 0 \\ 0 & -I_{r-q} \end{bmatrix} U$$

for some U that satisfies $UL = 0$.

[**10**] Show that $\Theta J \Theta^{\sim} = J$ for each J-unitary Θ.

[**11**] Obtain the dual representation of Problem [7].

Chapter 5

J-Lossless Conjugation and Interpolation

5.1 J-Lossless Conjugation

In this section, we introduce the notion of J-lossless conjugation which gives a powerful tool for computing (J, J')-lossless factorization. The conjugation is a simple operation of replacing a part of the poles of a transfer function by their "conjugates", that is, the mirror images with respect to the origin, by the multiplication of another transfer function. For instance, if a transfer function

$$G(s) = \frac{s+3}{(s+1)(s-2)} \tag{5.1}$$

is multiplied by an all-pass function $\Theta_-(s) = (s-2)/(s+2)$, then we have

$$G(s)\Theta_-(s) = \frac{s+3}{(s+1)(s-2)} \cdot \frac{s-2}{s+2} = \frac{s+3}{(s+1)(s+2)}.$$

Here, the pole at $s = 2$ of (5.1) is replaced by its conjugate $s = -2$ by the multiplication of $\Theta_-(s)$. On the other hand, the multiplication by $\Theta_+(s) = (s+1)/(s-1)$ yields

$$G(s)\Theta_+(s) = \frac{s+3}{(s+1)(s-2)} \cdot \frac{s+1}{s-1} = \frac{s+3}{(s-1)(s-2)}.$$

The pole at $s = -1$ in (5.1) is replaced by its conjugate $s = 1$. These are examples of conjugations. Since $\Theta_-(s)$ gives rise to a stable transfer function, it is called a *stabilizing conjugation*, whereas the $\Theta_+(s)$ is called an *anti-stabilizing conjugation.*

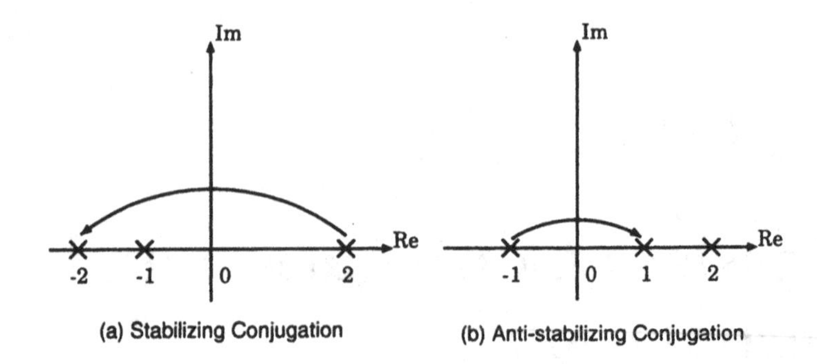

Figure 5.1 Illustration of J-Lossless Conjugations.

A special class of conjugations is important in H^∞ control theory, which is called a *J-lossless conjugation*, a conjugation by a J-lossless matrix.

DEFINITION 5.1 *A J-lossless matrix $\Theta(s)$ is said to be a* stabilizing *(*anti-stabilizing*)* J-lossless conjugator *of $G(s)$, if*

(i) $G(s)\Theta(s)$ is stable (anti-stable),

(ii) deg($\Theta(s)$) is equal to the number of unstable (stable) poles of $G(s)$, multiplicity included.

The property (i) implies that all the unstable poles of $G(s)$ must be cancelled out by the zeros of $\Theta(s)$. The property (ii) implies that the degree of $\Theta(s)$ is minimal for achieving (i). It rules out the cancellation of zeros of $G(s)$ by poles of $\Theta(s)$.

Let

$$G(s) = \left[\begin{array}{c|c} A & B \\ \hline C & D \end{array}\right] \in \mathbf{RL}^\infty_{(m+r)\times(p+r)} \tag{5.2}$$

be a state-space realization of $G(s)$. Remember the notation (4.51); that is,

$$J_{mr} := \begin{bmatrix} I_m & 0 \\ 0 & -I_r \end{bmatrix}.$$

The following result, which gives an existence condition for a J-lossless conjugator, plays an important role in H^∞ control theory.

THEOREM 5.2 *$G(s)$ given in (5.2) has a J-lossless stabilizing (anti-stabilizing) conjugator $\Theta(s)$ of $G(s)$ iff the Riccati equation*

$$XA + A^T X - XBJB^T X = 0 \tag{5.3}$$

has a solution $X \geq 0$ such that

$$\hat{A} := A - BJB^T X \tag{5.4}$$

is stable (anti-stable). In that case, a desired conjugator $\Theta(s)$ and the conjugated system $G(s)\Theta(s)$ are given, respectively, by

$$\Theta(s) = \left[\begin{array}{c|c} -A^T & XB \\ \hline -JB^T & I \end{array}\right] D_c = \left[\begin{array}{c|c} \hat{A} & B \\ \hline -JB^T X & I \end{array}\right] D_c, \tag{5.5}$$

$$G(s)\Theta(s) = \left[\begin{array}{c|c} \hat{A} & B \\ \hline C - DJB^T X & D \end{array}\right] D_c, \tag{5.6}$$

where D_c is any constant J-unitary matrix.

Proof. We only deal with the stabilizing case. The anti-stabilizing case can be treated analogously. We first assume that (A, B) is controllable. Suppose that there exists a stabilizing J-lossless conjugator $\Theta(s)$ with realization

$$\Theta(s) = \left[\begin{array}{c|c} A_c & B_c \\ \hline C_c & D_c \end{array}\right].$$

From the concatenation rule (2.13), it follows that

$$G(s)\Theta(s) = \left[\begin{array}{cc|c} A & BC_c & BD_c \\ 0 & A_c & B_c \\ \hline C & DC_c & DD_c \end{array}\right]. \tag{5.7}$$

We can find a similarity transformation of the A-matrix of (5.7) decomposing it into stable and anti-stable parts; that is,

$$\begin{bmatrix} A & BC_c \\ 0 & A_c \end{bmatrix} \begin{bmatrix} M_- & M_+ \\ N_- & N_+ \end{bmatrix} = \begin{bmatrix} M_- & M_+ \\ N_- & N_+ \end{bmatrix} \begin{bmatrix} A_- & 0 \\ 0 & A_+ \end{bmatrix}, \tag{5.8}$$

where A_- is stable and A_+ is anti-stable. Since (5.7) is stable, its anti-stable part must be uncontrollable. This implies that

$$\begin{bmatrix} BD_c \\ B_c \end{bmatrix} = \begin{bmatrix} M_- \\ N_- \end{bmatrix} B_1 \tag{5.9}$$

for some B_1. The matrix M_- is nonsingular. Indeed, if $\xi^T M_- = 0$ for some ξ, $\xi^T B = 0$ due to the first equation of (5.9). From (5.8), it follows that

$$AM_- + BC_cN_- = M_-A_-. \tag{5.10}$$

Hence, $\xi^T AM_- = 0$, which implies $\xi^T AB = 0$. Repeating this procedure, we can show that $\xi^T A^k B = 0$ for each k. Since (A, B) is controllable, $\xi = 0$. This implies that M_- is invertible.

Let $S := N_-M_-^{-1}$. From (5.9) and (5.10), it follows that

$$B_c = SBD_c, \quad A_- = M_-^{-1}(A + BC_cS)M_-. \tag{5.11}$$

Since Θ is J-lossless, Theorem 4.5 implies that

$$PA_c + A_c^T P + C_c^T JC_c = 0, \tag{5.12}$$

$$D_c^T JC_c + B_c^T P = 0 \tag{5.13}$$

for some $P \geq 0$. Due to (5.13) and (5.11), we have

$$C_c = -J(B_cD_c^{-1})^T P = -JB^T S^T P. \tag{5.14}$$

Also, from the second relation of (5.8), it follows that

$$\begin{aligned} A_cS &= A_cN_-M_-^{-1} = N_-A_-M_-^{-1} = S(A + BC_cS) \\ &= S(A - BJB^T X), \end{aligned} \tag{5.15}$$

where $X := S^T PS$. Also, from (5.12), it follows that

$$S^T PA_cS + S^T A_c^T PS + S^T C_c^T JC_cS = 0. \tag{5.16}$$

Substitution of (5.14) and (5.15) in (5.16) yields (5.3). Due to (5.11), $\hat{A} = A - BJB^T X = A + BC_cS = M_-A_-M_-^{-1}$. Therefore, $\hat{A}$ is stable. Thus the first assertion has been proven.

From (5.11) and (5.14), it follows that

$$\Theta = \left[\begin{array}{c|c} A_c & B_c \\ \hline C_c & D_c \end{array}\right] = \left[\begin{array}{c|c} A_c & SB \\ \hline -JB^T S^T P & I \end{array}\right] D_c.$$

Due to (5.15), $(sI - A_c)^{-1}S = S(sI - \hat{A})^{-1}$. It follows that

$$\Theta = \left[\begin{array}{c|c} \hat{A} & B \\ \hline -JB^T X & I \end{array}\right] D_c = \left[\begin{array}{c|c} -A^T & XB \\ \hline -JB^T & I \end{array}\right] D_c,$$

which verifies (5.5). The second equality is due to $X(sI-\hat{A})^{-1}=(sI+A^T)^{-1}X$.
From the concatenation rule, it follows that

$$G\Theta = \left[\begin{array}{cc|c} A & -BJB^T & B \\ 0 & -A^T & XB \\ \hline C & -DJB^T & D \end{array}\right] D_c.$$

Taking the similarity transformation with

$$T=\begin{bmatrix} I & 0 \\ -X & I \end{bmatrix}, \quad T^{-1}=\begin{bmatrix} I & 0 \\ X & I \end{bmatrix}$$

yields (5.6).

It remains to show that $\deg(\Theta)$ is equal to the number of unstable poles of $G(s)$. To see this, we note that (5.3) implies

$$X\hat{A}+A^TX=0. \tag{5.17}$$

Let ξ be an eigenvector of A corresponding to a stable eigenvalue λ; that is, $A\xi=\lambda\xi$. Then, from (5.17), it follows that $\xi^TX(\lambda I+\hat{A})=0$. Since $\hat{A}$ is stable, we have $\xi^TX=0$. From the state-space form (5.5) of Θ, we conclude that λ is an uncontrollable mode of Θ. Thus the assertion has been established.

Finally, we relax the controllability assumption. If (A,B) is not stabilizable, $\hat{A}$ given in (5.4) cannot be stable. Hence we can assume (A,B) is stabilizable. In that case, we can find a similarity transformation such that (A,B) is represented as

$$A=\begin{bmatrix} A_{11} & 0 & 0 \\ 0 & A_{22} & A_{23} \\ 0 & 0 & A_{33} \end{bmatrix}, \quad B=\begin{bmatrix} B_1 \\ B_2 \\ 0 \end{bmatrix}, \tag{5.18}$$

where A_{11} is anti-stable, A_{22} and A_{33} are stable and the pair

$$\left(\begin{bmatrix} A_{11} & 0 \\ 0 & A_{22} \end{bmatrix}, \begin{bmatrix} B_1 \\ B_2 \end{bmatrix}\right) \tag{5.19}$$

is controllable. If $G(s)$ has a J-lossless conjugator, there exists a $\bar{X}\geq 0$ satisfying (5.3) for the pair (5.19). It is easy to see that $\bar{X}$ is of the form

$$\bar{X}=\begin{bmatrix} X_1 & 0 \\ 0 & 0 \end{bmatrix},$$

where $X_1 \geq 0$ satisfies (5.3) for (A_{11}, B_1). It is now obvious that

$$X = \begin{bmatrix} X_1 & 0 & 0 \\ 0 & 0 & 0 \\ 0 & 0 & 0 \end{bmatrix}$$

gives a solution to (5.3) for (A, B). The proof is now complete. ∎

Remark: The matrix X in Theorem 5.2 is written as

$$X = Ric\left(\begin{bmatrix} A & -BJB^T \\ 0 & -A^T \end{bmatrix}\right).$$

Example 5.1
Let

$$G(s) = \begin{bmatrix} \dfrac{1}{s-2} & \dfrac{\alpha}{s-2} \\ 0 & \dfrac{1}{s+1} \end{bmatrix} = \left[\begin{array}{cc|cc} 2 & 0 & 1 & \alpha \\ 0 & -1 & 0 & 1 \\ \hline 1 & 0 & 0 & 0 \\ 0 & 1 & 0 & 0 \end{array}\right],$$

where α is a real parameter. For $J = J_{1,1}$, the Riccati equation (5.3) is given in this case by

$$X \begin{bmatrix} 2 & 0 \\ 0 & -1 \end{bmatrix} + \begin{bmatrix} 2 & 0 \\ 0 & -1 \end{bmatrix} X - X \begin{bmatrix} 1-\alpha^2 & -\alpha \\ -\alpha & -1 \end{bmatrix} X = 0.$$

The stabilizing solution is given by

$$X = \begin{bmatrix} \delta & 0 \\ 0 & 0 \end{bmatrix}, \quad \delta := \frac{4}{1-\alpha^2}$$

$$\hat{A} = \begin{bmatrix} -2 & 0 \\ \delta\alpha & -1 \end{bmatrix}.$$

Therefore, $G(s)$ has a stabilizing $J_{1,1}$-lossless conjugator iff $|\alpha| < 1$. In that case, the conjugator (5.5) is given by

$$\Theta(s) = \left[\begin{array}{cc|cc} -2 & 0 & \delta & \delta\alpha \\ 0 & 1 & 0 & 0 \\ \hline -1 & 0 & 1 & 0 \\ \alpha & 1 & 0 & 1 \end{array}\right] = \left[\begin{array}{c|cc} -2 & \delta & \delta\alpha \\ \hline -1 & 1 & 0 \\ \alpha & 0 & 1 \end{array}\right],$$

where we chose $D_c = I$. The conjugated system is given by

$$G(s)\Theta(s) = \left[\begin{array}{cc|cc} -2 & 0 & 1 & \alpha \\ \delta\alpha & -1 & 0 & 1 \\ \hline 1 & 0 & 0 & 0 \\ 0 & 1 & 0 & 0 \end{array}\right].$$

Since the conjugator Θ depends only on (A, B) as seen from the form (5.5), Θ is sometimes called a *J-lossless conjugator of* (A, B).

All the zeros of the conjugator are supplied just for cancelling out the poles of $G(s)$ that are to be conjugated. It does not create any new zero. This reflects the minimality of Definition 5.1. This property is stated in a more precise way as follows.

LEMMA 5.3 *If $\Theta(s)$ is a J-lossless conjugator of $G(s)$, each zero of the conjugated system $G(s)\Theta(s)$ is a zero of $G(s)$.*

Proof. Let λ be a zero of $G(s)\Theta(s)$ whose state-space realization is given by (5.6). There exist nonzero vectors ξ and η such that

$$(A - BJB^T X - \lambda I)\xi + BD_c\eta = 0,$$
$$(C - DJB^T X)\xi + DD_c\eta = 0.$$

Taking $\bar{\eta} := D_c\eta - JB^T X\xi$ yields

$$(A - \lambda I)\xi + B\bar{\eta} = 0, \quad C\xi + D\bar{\eta} = 0,$$

which implies that λ is a zero of $G(s)$. ∎

Lemma 5.3 implies that each zero of a conjugator is to be cancelled out by a pole of the system to be conjugated. Due to Lemma 4.7, we have the following interesting property of the J-lossless conjugator.

LEMMA 5.4 *Let λ be a pole of $G(s)$ to be conjugated by a J-lossless conjugator $\Theta(s)$. Then $\Theta(s)$ must have $-\lambda$ as its pole.*

We conclude this section stating the following two properties of the Riccati equation (5.3).

LEMMA 5.5 *If A has an eigenvalue on the $j\omega$-axis, then the Riccati equation (5.3) has no stabilizing solution.*

Proof. Let $j\omega$ be an eigenvalue of A with the eigenvector $\xi \neq 0$ and assume that Equation (5.3) has a solution X such that $\hat{A}$ in (5.4) is stable. Since $A\xi = j\omega\xi$, Equation (5.3) implies that $\xi^T X(j\omega I + \hat{A}) = 0$. Since $\hat{A}$ is stable, $X\xi = 0$. Then, $\hat{A}\xi = A\xi = j\omega\xi$, which contradicts the assumption that $\hat{A}$ is stable. ∎

LEMMA 5.6 *If A is anti-stable ($-A$ is stable) and the pair (A, B) has a stabilizing J-lossless conjugator, the stabilizing solution X of (5.3) is invertible.*

Proof. Assume, contrary to the assertion, that the solution X of (5.3) that stabilizes (5.4) has a nonempty kernel. Let $\xi \in \mathrm{Ker}X$. From (5.3), it follows that $XA\xi = 0$. Hence, Ker$X$ is A-invariant. Therefore, there exists an eigenvector $\xi_0 \in \mathrm{Ker}X$ of A corresponding to an eigenvalue λ_0; that is, $A\xi_0 = \lambda_0\xi_0$, $X\xi_0 = 0$. This implies that $\hat{A}\xi_0 = \lambda_0\xi_0$; that is, λ_0 is also an eigenvalue of $\hat{A}$. But this is impossible because $\hat{A}$ is stable and A is anti-stable. Thus Ker$X$ is void. ∎

5.2 Connections to Classical Interpolation Problem

In this section, we discuss the relation between the J-lossless conjugation and the well-known Nevanlinna-Pick interpolation problem. The Nevanlinna-Pick problem (NP problem, for short) is quite simple in its formulation but it exhibits the deep structure of analytic functions and played a fundamental role in the development of classical analysis and operator theory. It is shown in this section that the J-lossless conjugation introduced in the preceding section solves the NP problem in the state space.

Now we formulate the NP problem.

[Nevanlinna-Pick Interpolation Problem]

Let $(\alpha_1, \beta_1), (\alpha_2, \beta_2), \cdots, (\alpha_p, \beta_p)$ be a set of p pairs of complex numbers such that

$$\mathrm{Re}\ \alpha_i > 0, \quad |\beta_i| < 1, \quad i = 1, 2, \cdots, p.$$

Find a function $f(s) \in \mathbf{BH}^\infty$ that satisfies the interpolation conditions

$$f(\alpha_i) = \beta_i, \quad i = 1, 2, \cdots, p. \tag{5.20}$$

The solution to the NP problem was initially given by Pick [78]:

THEOREM 5.7 *The NP problem is solvable if and only if the matrix P given by*

$$P := \begin{bmatrix} \dfrac{1-\beta_1\bar{\beta}_1}{\alpha_1+\bar{\alpha}_1} & \cdots & \dfrac{1-\beta_1\bar{\beta}_p}{\alpha_1+\bar{\alpha}_p} \\ & \cdots\cdots & \\ \dfrac{1-\beta_p\bar{\beta}_1}{\alpha_p+\bar{\alpha}_1} & \cdots & \dfrac{1-\beta_p\bar{\beta}_p}{\alpha_p+\bar{\alpha}_p} \end{bmatrix} \tag{5.21}$$

is positive definite.

Pick [78] proved the necessity part of Theorem 5.7 based on the Schwarz Lemma in classical analysis. Nevanlinna [77] proved the sufficiency part by actually constructing a solution recursively. We show that the NP problem is just a special case of J-lossless conjugation.

THEOREM 5.8 *The NP problem is solvable iff the pair (A, B) given by*

$$A = \begin{bmatrix} \alpha_1 & 0 & \cdots & 0 \\ 0 & \alpha_2 & \cdots & 0 \\ & \cdots\cdots & & \\ 0 & 0 & \cdots & \alpha_p \end{bmatrix}, \quad B = \begin{bmatrix} 1 & -\beta_1 \\ 1 & -\beta_2 \\ & \cdots \\ 1 & -\beta_p \end{bmatrix} \tag{5.22}$$

has a stabilizing J-lossless conjugator Θ for

$$J = \begin{bmatrix} 1 & 0 \\ 0 & -1 \end{bmatrix}. \tag{5.23}$$

In that case,

$$f = HM\,(\Theta; u)$$

solves the problem for any $u \in \mathbf{BH}_\infty$.

Proof. To prove the sufficiency, assume that Θ is a stabilizing J-lossless conjugator of (A, B). From the definition,

$$(sI - A)^{-1}B\Theta(s) = H(s)$$

is stable. Multiplying both sides of this relation by $s - \alpha_i$ and letting $s \to \alpha_i$ yields

$$\begin{bmatrix} 1 & -\beta_i \end{bmatrix} \Theta(\alpha_i) = 0,$$

which implies

$$\theta_{11}(\alpha_i) = \beta_i \theta_{21}(\alpha_i), \quad \theta_{12}(\alpha_i) = \beta_i \theta_{22}(\alpha_i).$$

Since $f = HM\ (\Theta; u) = (\theta_{11}u + \theta_{12})/(\theta_{21}u + \theta_{22})$, we see that f satisfies the interpolation conditions (5.20). Due to Theorem 4.15, $f \in \mathbf{BH}^\infty$, for each u. The proof of sufficiency is now complete.

We prove the necessity by induction with respect to the number of interpolation constraints p. Assume that $f(s)$ is a solution to the NP problem. Let

$$\Theta_1(s) := \frac{1}{\sqrt{1-|\beta_1|^2}} \begin{bmatrix} B_1(s) & \beta_1 \\ \bar{\beta}_1 B_1(s) & 1 \end{bmatrix}, \tag{5.24}$$

where

$$B_1(s) := \frac{s-\alpha_1}{s+\bar{\alpha}_1}.$$

Since $B_1(s)$ is inner, that is, $B_1^\sim(s) B_1(s) = 1$, $\Theta_1(s)$ given by (5.24) is J-lossless with J being given in (5.23). Direct manipulation yields

$$\frac{1}{s-\alpha_1} \begin{bmatrix} 1 & -\beta_1 \end{bmatrix} \Theta_1(s) = \sqrt{1-|\beta_1|^2} \begin{bmatrix} \dfrac{1}{s+\bar{\alpha}_1} & 0 \end{bmatrix}.$$

Thus, $\Theta_1(s)$ is a stabilizing J-lossless conjugator of

$$G_1(s) = \frac{1}{s-\alpha_1} \begin{bmatrix} 1 & -\beta_1 \end{bmatrix}.$$

The assertion has now been proven for $p = 1$.

Assume that the assertion holds for $p-1$, and define

$$f_1 := HM\ (\Theta_1^{-1}; f).$$

Since

$$\Theta_1^{-1}(s) = \frac{1}{\sqrt{1-|\beta_1|^2}} \begin{bmatrix} \dfrac{s+\bar{\alpha}_1}{s-\alpha_1} & -\dfrac{s+\bar{\alpha}_1}{s-\alpha_1}\beta_1 \\ -\bar{\beta}_1 & 1 \end{bmatrix},$$

$f_1(s)$ is explicitly given by

$$f_1(s) = \frac{s+\bar{\alpha}_1}{s-\alpha_1} \cdot \frac{f(s)-\beta_1}{1-\bar{\beta}_1 f(s)}.$$

Since $f(\alpha_1) = \beta_1$, the unstable pole at $s = \alpha_1$ is cancelled. Since $|\bar{\beta}_1 f(s)| < 1$ for $\mathrm{Re}[s] \geq 0$, $f_1(s)$ is stable. Due to the identity

$$|1 - \bar{\beta}_1 f(s)|^2 = |f(s) - \beta_1|^2 + (1 - |\beta_1|^2) \cdot (1 - |f(s)|^2),$$

we see that

$$1 - |f_1(j\omega)|^2 = \frac{(1 - |\beta_1|^2)(1 - |f(j\omega)|^2)}{|1 - \bar{\beta}_1 f(j\omega)|^2} \geq 0.$$

Therefore, $f_1 \in \mathbf{BH}_\infty$. From the interpolation condition (5.20), it follows that

$$f_1(\alpha_i) = \beta_i', \quad i = 2, 3, \cdots, p. \qquad (5.25)$$

$$\beta_i' := \frac{\alpha_i + \bar{\alpha}_1}{\alpha_i - \alpha_1} \cdot \frac{\beta_i - \beta_1}{1 - \bar{\beta}_1 \beta_i}.$$

Since the assertion holds for $p - 1$, we have a stabilizing J-lossless conjugator $\Theta'(s)$ for (A', B') given by

$$A' = \begin{bmatrix} \alpha_2 & 0 & \cdots & 0 \\ 0 & \alpha_3 & \cdots & 0 \\ \vdots & & \ddots & \vdots \\ 0 & 0 & \cdots & \alpha_p \end{bmatrix}, \quad B' = \begin{bmatrix} 1 & -\beta_2' \\ 1 & -\beta_3' \\ & \cdots \\ 1 & -\beta_p' \end{bmatrix}, \qquad (5.26)$$

due to the solvability of the interpolation problem (5.25). Direct manipulations yield

$$\frac{1}{s - \alpha_i} \begin{bmatrix} 1 & -\beta_i \end{bmatrix} \Theta_1(s) = \begin{bmatrix} \dfrac{\nu_i}{s + \bar{\alpha}_1} & 0 \end{bmatrix} + \frac{\mu_i}{s - \alpha_i} \begin{bmatrix} 1 & -\beta_i' \end{bmatrix}, \qquad (5.27)$$

where

$$\nu_i := \frac{1 - \bar{\beta}_1 \beta_i}{\sqrt{1 - |\beta_1|^2}} \cdot \frac{\alpha_1 + \bar{\alpha}_1}{\alpha_i + \bar{\alpha}_1},$$

$$\mu_i := \frac{1 - \bar{\beta}_1 \beta_i}{\sqrt{1 - |\beta_1|^2}} \cdot \frac{\alpha_i - \alpha_1}{\alpha_i + \bar{\alpha}_1}.$$

Thus,

$$(sI - A)^{-1}B\Theta_1(s) = N\begin{bmatrix} \frac{1}{s+\bar{\alpha}_1} & 0 \end{bmatrix} + M(sI - A')^{-1}B'$$

$$N = \begin{bmatrix} \nu_1 & \nu_2 & \cdots & \nu_p \end{bmatrix}^T, \quad M = \begin{bmatrix} 0 & 0 & \cdots & 0 \\ \mu_2 & 0 & \cdots & 0 \\ 0 & \mu_3 & \cdots & 0 \\ \vdots & \vdots & \ddots & \vdots \\ 0 & 0 & \cdots & \mu_p \end{bmatrix}.$$

Since $(sI - A')^{-1}B'\Theta'(s)$ is stable from the definition of $\Theta'(s)$, we see that $(sI - A)^{-1}B\Theta_1(s)\Theta'(s)$ is stable. Thus, we have proven that

$$\Theta(s) := \Theta_1(s)\Theta'(s)$$

is a stabilizing J-lossless conjugator of (A, B) given by (5.22). ∎

Since A given in (5.22) is anti-stable, Lemma 5.6 implies that the solution X of the Riccati equation (5.3) for (A, B) given in (5.22) has the inverse $P = X^{-1}$ which satisfies a Lyapunov type equation

$$AP + PA^* = BJB^*. \tag{5.28}$$

Here A^T and B^T are replaced by their Hermitian conjugate A^* and B^* in (5.3). It is straightforward to see that the solution P of Equation (5.28) is identical to the Pick matrix (5.21). Therefore, Theorem 5.8 is another expression of the classical result Theorem 5.7.

5.3 Sequential Structure of J-Lossless Conjugation

In the proof of the necessity part of Theorem 5.8, we introduced in (5.24) a J-lossless matrix

$$\Theta_1(s) = \frac{1}{\sqrt{1-|\beta_1|^2}}\begin{bmatrix} B_1(s) & \beta_1 \\ \bar{\beta}_1 B_1(s) & 1 \end{bmatrix}. \tag{5.29}$$

In the state space, it is represented as

$$\Theta_1(s) = \left[\begin{array}{c|cc} -\bar{\alpha}_1 & x_1 & -x_1\beta_1 \\ \hline -1 & 1 & 0 \\ -\bar{\beta}_1 & 0 & 1 \end{array}\right] R_1, \tag{5.30}$$

where

$$x_1 := \frac{\alpha_1 + \bar{\alpha}_1}{1 - |\beta_1|^2}, \tag{5.31}$$

$$R_1 = \frac{1}{\sqrt{1 - |\beta_1|^2}} \begin{bmatrix} 1 & \beta_1 \\ \bar{\beta}_1 & 1 \end{bmatrix}. \tag{5.32}$$

Obviously, x_1 is a solution to the Riccati equation (5.3) for the pair $(\alpha_1, [1 \ \ -\beta_1])$. Therefore, $\Theta_1(s)$ is of the form (5.5) for the pair $(\alpha_1, [1 \ \ -\beta_1])$ with D_c being given by R_1 in (5.32). Thus, $\Theta_1(s)$ is a stabilizing J-lossless conjugator of $(\alpha_1, [1 \ \ -\beta_1])$. As is seen in (5.27), the conjugation by $\Theta_1(s)$ gives rise to another interpolation problem given by (A', B') in (5.26) which has $p-1$ interpolation constraints compared with the p constraints in the original interpolation problem. Next, we can find a stabilizing J-lossless conjugator $\Theta_2(s)$ for $(\alpha_2, [1 \ \ -\beta_2'])$ that gives rise to a new interpolation problem having $p-2$ interpolation constraints. In this way, we can reduce the number of interpolation constraints one by one and finally reach the stage where all the interpolation constraints are exhausted. At this stage, we can choose an arbitrary function in $\mathbf{BH}^\infty$ to complete the parameterization of the set of solutions of the NP problem. This is the algorithm established by Nevanlinna, which is actually equivalent to computing the Cholesky decomposition of the Pick matrix.

A block diagram of $\Theta_1(s)$ is shown in Figure 5.2 which represents the celebrated *lattice form* [18]. It is known that every $J_{1,1}$-lossless system of degree 1 is always of the form (5.29).

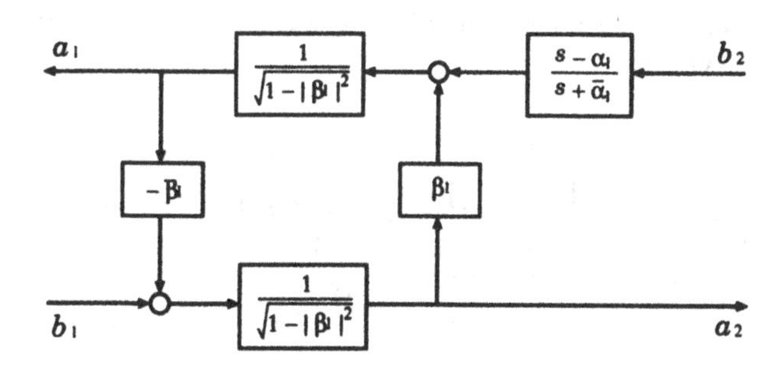

Figure 5.2 Lattice Realization of $\Theta_1(s)$ in (5.29).

Now we show that the aforementioned sequential structure that was extensively investigated by Potapov [79] is intrinsic to the J-lossless conjugation. We confine ourselves to the case where $G(s)$ is anti-stable; that is, $-A$ is stable. Now assume that $G(s)$ in (5.2) is of the form

$$G(s) = \left[\begin{array}{cc|c} A_{11} & 0 & B_1 \\ A_{21} & A_{22} & B_2 \\ \hline C_1 & C_2 & D \end{array}\right]. \tag{5.33}$$

Assume that there exists a stabilizing J-lossless conjugator of $G(s)$. Then, due to Lemma 5.6, there exists a matrix $P > 0$ that solves the Lyapunov equation (5.28) represented in this case as

$$\begin{aligned} \begin{bmatrix} A_{11} & 0 \\ A_{21} & A_{22} \end{bmatrix} \begin{bmatrix} P_{11} & P_{12} \\ P_{12}^T & P_{22} \end{bmatrix} + \begin{bmatrix} P_{11} & P_{12} \\ P_{12}^T & P_{22} \end{bmatrix} \begin{bmatrix} A_{11}^T & A_{21}^T \\ 0 & A_{22}^T \end{bmatrix} \\ = \begin{bmatrix} B_1 \\ B_2 \end{bmatrix} J \begin{bmatrix} B_1^T & B_2^T \end{bmatrix}. \end{aligned} \tag{5.34}$$

Therefore, we have

$$A_{11}P_{11} + P_{11}A_{11}^T = B_1JB_1^T.$$

Since $-A_{11}^T$ is stable, $A_{11} - B_1JB_1^TP_{11}^{-1} = -P_{11}A_{11}^TP_{11}^{-1}$ is stable. Hence (A_{11}, B_1) has a stabilizing J-lossless conjugator given by

$$\Theta_1(s) = \left[\begin{array}{c|c} -A_{11}^T & P_{11}^{-1}B_1 \\ \hline -JB_1^T & I \end{array}\right] D_{c_1}, \tag{5.35}$$

where D_{c_1} is an arbitrary constant J-lossless matrix. Straightforward manipulations yield

$$G(s)\Theta_1(s) = \left[\begin{array}{ccc|c} A_{11} & 0 & -B_1JB_1^T & B_1 \\ A_{21} & A_{22} & -B_2JB_1^T & B_2 \\ 0 & 0 & -A_{11}^T & P_{11}^{-1}B_1 \\ \hline C_1 & C_2 & -DJB_1^T & D \end{array}\right] D_{c_1}.$$

The similarity transformation with the matrix

$$T = \begin{bmatrix} I & 0 & -P_{11} \\ 0 & I & -P_{12}^T \\ 0 & 0 & I \end{bmatrix}, \quad T^{-1} = \begin{bmatrix} I & 0 & P_{11} \\ 0 & I & P_{12}^T \\ 0 & 0 & I \end{bmatrix}$$

and cancelling out the uncontrollable portion yield

$$G(s)\Theta_1(s) = \left[\begin{array}{cc|c} -A_{11}^T & 0 & P_{11}^{-1}B_1 \\ 0 & A_{22} & B_2 - P_{12}^T P_{11}^{-1} B_1 \\ \hline \hat{C}_1 & C_2 & D \end{array}\right] D_{c_1},$$

where

$$\hat{C}_1 := C_1 P_{11} + C_2 P_{12}^T - DJB_1^T.$$

Now we conjugate the remaining unstable portion

$$G_2 := \left[\begin{array}{c|c} A_{22} & B_2 - P_{12}^T P_{11}^{-1} B_1 \\ \hline C_2 & D \end{array}\right] D_{c_1} \tag{5.36}$$

of $G(s)\Theta(s)$. From Equation (5.34), we can show that the Schur complement

$$\hat{P} := P_{22} - P_{12}^T P_{11}^{-1} P_{12}$$

of P satisfies

$$A_{22}\hat{P} + \hat{P}A_{22}^T = (B_2 - P_{12}^T P_{11}^{-1} B_1) J (B_2^T - B_1^T P_{11}^{-1} P_{12}),$$

which corresponds to (5.28) for G_2 in (5.36). This implies that $G_2(s)$ has a stabilizing J-lossless conjugator given by

$$\Theta_2 := \left[\begin{array}{c|c} -A_{22}^T & \hat{P}^{-1}(B_2 - P_{12}^T P_{11}^{-1} B_1) \\ \hline J(B_1^T P_{11}^{-1} P_{12} - B_2^T) & I \end{array}\right] D_{c2}. \tag{5.37}$$

Now the concatenated system is calculated to be

$$\Theta_1(s)\Theta_2(s) = \left[\begin{array}{cc|c} -A_{11}^T & U & P_{11}^{-1}B_1 \\ 0 & -A_{22}^T & \hat{P}^{-1}\hat{B} \\ \hline -JB_1^T & -J\hat{B}^T & I \end{array}\right], \tag{5.38}$$

where

$$U = -P_{11}^{-1} B_1 J \hat{B}^T,$$
$$\hat{B} = B_2 - P_{12}^T P_{11}^{-1} B_1,$$

and we take $D_{c1} = D_{c2} = I$. Taking the similarity transformation of (5.38) with the transformation matrix

$$T = \begin{bmatrix} I & -P_{11}^{-1}P_{12} \\ 0 & I \end{bmatrix}, \quad T^{-1} = \begin{bmatrix} I & P_{11}^{-1}P_{12} \\ 0 & I \end{bmatrix}$$

and using the identity

$$\begin{bmatrix} P_{11} & P_{12} \\ P_{12}^T & P_{22} \end{bmatrix}^{-1} = \begin{bmatrix} (I + P_{11}^{-1}P_{12}\hat{P}^{-1}P_{12}^T)P_{11}^{-1} & -P_{11}^{-1}P_{12}\hat{P}^{-1} \\ -\hat{P}^{-1}P_{12}^T P_{11}^{-1} & \hat{P}^{-1} \end{bmatrix},$$

we can see that

$$\Theta_1(s)\Theta_2(s) = \left[\begin{array}{c|c} -A^T & P^{-1}B \\ \hline -JB^T & I \end{array}\right].$$

The system on the right-hand side is identical to the stabilizing J-lossless conjugator of $G(s)$ given in Theorem 5.2. Thus, we have shown that the stabilizing J-lossless conjugator $\Theta(s)$ of $G(s)$ given in (5.33) is factored into the product of two J-lossless conjugators $\Theta_1(s)$ of (5.35) and $\Theta_2(s)$ of (5.37), each of which conjugates a portion of $G(s)$ according to the polar decomposition of the A-matrix of $G(s)$ in the state-space form (5.33). This result is easily generalized as follows.

THEOREM 5.9 *Let $G(s)$ be an anti-stable system whose state-space realization is given by*

$$G(s) = \left[\begin{array}{cccc|c} A_{11} & 0 & \cdots & 0 & B_1 \\ A_{21} & A_{22} & \cdots & 0 & B_2 \\ \vdots & & \ddots & \vdots & \vdots \\ A_{m1} & A_{m2} & \cdots & A_{mm} & B_m \\ \hline C_1 & C_2 & \cdots & C_m & D \end{array}\right].$$

If $G(s)$ has a stabilizing J-lossless conjugator $\Theta(s)$, then it can be represented as a product of m J-lossless systems

$$\Theta(s) = \Theta_1(s)\Theta_2(s)\cdots\Theta_m(s), \tag{5.39}$$

such that a state-space form of $\Theta_i(s)$ has $-A_{ii}^T$ as its A-matrix.

If A_{ii} in the polar decomposition $G(s)$ is scalar, we can realize its conjugator as cascade connections of the lattice form of Figure 5.2 as in Figure 5.3.

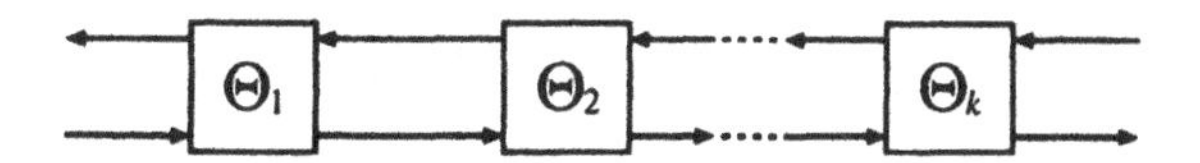

Figure 5.3 Cascade Realization of J-Lossless Conjugator.

Example 5.2
Let us conjugate sequentially the pair (A, B) given by

$$\left(\begin{bmatrix} 1 & 0 \\ 0 & 2 \end{bmatrix}, \begin{bmatrix} 1 & 1/2 \\ 1 & 1/3 \end{bmatrix} \right). \tag{5.40}$$

First, conjugate the pair $(A_{11}, B_1) = (1, [\ 1 \quad 1/2\ ])$. The solution to the Riccati equation (5.28) in this case is given by

$$P_{11} = \frac{3}{8}$$

and the conjugator is given by

$$\Theta_1(s) = \left[\begin{array}{c|cc} -1 & 8/3 & 4/3 \\ \hline -1 & 1 & 0 \\ 1/2 & 0 & 1 \end{array} \right] D_{c_1}. \tag{5.41}$$

The conjugated system is calculated to be

$$(sI - A)^{-1} B \Theta_1(s) = \left[\begin{array}{ccc|cc} 1 & 0 & -3/4 & 1 & 1/2 \\ 0 & 2 & -5/6 & 1 & 1/3 \\ 0 & 0 & -1 & 8/3 & 4/3 \\ \hline 1 & 0 & 0 & 0 & 0 \\ 0 & 1 & 0 & 0 & 0 \end{array} \right] D_{c_1}.$$

Taking the similarity transformation with

$$T = \begin{bmatrix} 1 & 0 & -3/8 \\ 0 & 1 & -5/18 \\ 0 & 0 & 1 \end{bmatrix}, \quad T^{-1} = \begin{bmatrix} 1 & 0 & 3/8 \\ 0 & 1 & 5/18 \\ 0 & 0 & 1 \end{bmatrix},$$

and omitting the uncontrollable portion, we obtain

$$(sI-A)^{-1}B\Theta_1(s)=\left[\begin{array}{cc|cc} -1 & 0 & 8/3 & 4/3 \\ 0 & 2 & 7/27 & -1/27 \\ \hline 3/8 & 0 & 0 & 0 \\ 5/18 & 1 & 0 & 0 \end{array}\right]D_{c_1}.$$

Now the remaining portion $(2, [\ 7/27 \quad -1/27\ ])$ is to be conjugated. The corresponding Riccati equation is solved to be

$$\hat{X}_{22}=\frac{4}{243} \tag{5.42}$$

and the J-lossless conjugator is given by

$$\Theta_2(s)=\left[\begin{array}{c|cc} -2 & 63/4 & -9/4 \\ \hline -7/27 & 1 & 0 \\ -1/27 & 0 & 1 \end{array}\right]D_{c_2}. \tag{5.43}$$

Now the concatenation of $\Theta_1(s)$ and $\Theta_2(s)$ yields

$$\Theta_1(s)\Theta_2(s)=\left[\begin{array}{cc|cc} -1 & -20/27 & 8/3 & 4/3 \\ 0 & -2 & 63/4 & -9/4 \\ \hline -1 & -7/27 & 1 & 0 \\ 1/2 & -1/27 & 0 & 1 \end{array}\right], \tag{5.44}$$

where we took $D_{c_1}=D_{c_2}=I$ for brevity. On the other hand, the $J_{1,1}$-lossless conjugator of the original system (5.40) is calculated to be

$$\Theta(s)=\left[\begin{array}{cc|cc} -1 & 0 & \multicolumn{2}{c}{\multirow{2}{*}{$P^{-1}\begin{bmatrix} 1 & 1/2 \\ 1 & 1/3 \end{bmatrix}$}} \\ 0 & -2 & & \\ \hline -1 & -1 & 1 & 0 \\ 1/2 & 1/3 & 0 & 1 \end{array}\right], \tag{5.45}$$

where P is the solution to

$$\begin{bmatrix} 1 & 0 \\ 0 & 2 \end{bmatrix}P+P\begin{bmatrix} 1 & 0 \\ 0 & 2 \end{bmatrix}=\begin{bmatrix} 1 & 1/2 \\ 1 & 1/3 \end{bmatrix}\begin{bmatrix} 1 & 1 \\ -1/2 & -1/3 \end{bmatrix}.$$

Actually, P and P^{-1} are given, respectively, by

$$P = \begin{bmatrix} 3/8 & 5/18 \\ 5/18 & 2/9 \end{bmatrix}, \quad P^{-1} = \begin{bmatrix} 36 & -45 \\ -45 & 243/4 \end{bmatrix}, \tag{5.46}$$

which are positive definite. Taking the similarity transformation with

$$T = \begin{bmatrix} 1 & -20/27 \\ 0 & 1 \end{bmatrix}, \quad T^{-1} = \begin{bmatrix} 1 & 20/27 \\ 0 & 1 \end{bmatrix}$$

to (5.44), we easily see that the product $\Theta_1(s)\Theta_2(s)$ is identical to $\Theta(s)$ in (5.45). Also, note that $\hat{P}_{22}$ given in (5.42) is identical to the Schur complement of P given in (5.46). Taking D_{c_1} in (5.41) as in (5.30) and (5.32) where $x_1 = 8/3$ and $\beta_1 = -1/2$, we can construct a lattice realization of $\Theta_1(s)$ as in Figure 5.2. Also, $\Theta_2(s)$ given in (5.43) is represented as

$$\Theta_2(s) = \left[\begin{array}{c|cc} -2 & x_2 & -x_2\beta_2 \\ \hline -1 & 1 & 0 \\ -\beta_2 & 0 & 1 \end{array}\right] R_2,$$

$$R_2 = \frac{1}{\sqrt{1-|\beta_2|^2}} \begin{bmatrix} 1 & \beta_2 \\ \beta_2 & 1 \end{bmatrix},$$

where $x_2 = 49/12$, $\beta_2 = 1/7$, and $D_{c_2} = R_2$. Hence, we can obtain the lattice realization of $\Theta_2(s)$. Figure 5.4 shows a lattice realization of the J-lossless conjugator of (5.40).

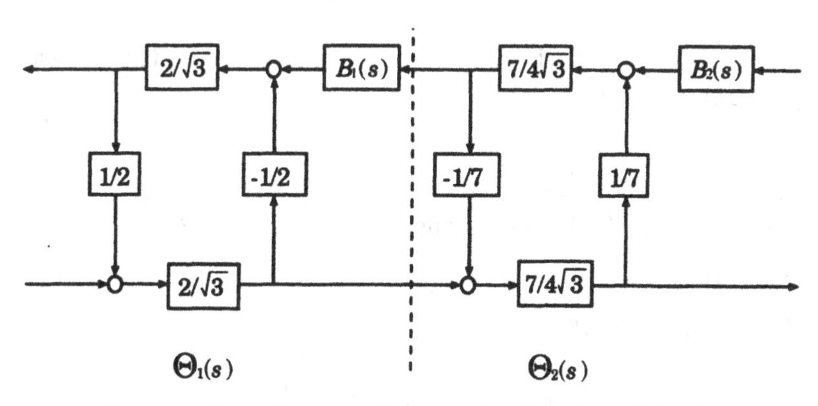

Figure 5.4 Cascade Lattice Realization of $\Theta(s)$.

Notes

The subject of this chapter was J-lossless conjugation. In the earlier version of J-lossless conjugation which was initially proposed in [54], the characterization was given in terms of X^{-1} in Equation (5.3) rather than X. The general case stated in Theorem 5.2 was obtained recently in [59]. The relationship between the J-lossless conjugation and the classical interpolation was extensively investigated in [55]. There, it was shown that the J-lossless conjugation is nothing but a state-space representation of the Nevanlinna-Pick interpolation theory [56]. The discrete-time version is found in [64]. Its relation to lattice realization which was an important research field in digital signal processing [51][80] was also discussed in [56]. The J-lossless conjugation is also regarded as a state-space representation of the cascade synthesis of passive electrical circuits [38][104] which dates back to Darlington Synthesis [13]. This was initially pointed out in [20]. Further elaboration of this subject will be the subject of future work.

Problems

[**1**] Show that stabilizing J-lossless conjugators are stable.

[**2**] Prove that an anti-stabilizing J-lossless conjugator for (A, B) exists, iff

$$X = Ric\left(\begin{bmatrix} -A & BJB^T \\ 0 & A^T \end{bmatrix}\right)$$

exists.

[**3**] Show that, for an arbitrary $\alpha_1 \in \mathbf{C}^+$ and $\beta_1 \in \mathbf{D}$, there exists a $f \in \mathbf{BH}_\infty$ such that

$$f(\alpha_1) = \beta_1. \tag{5.47}$$

[**4**] Show that, if (A, B) has a stabilizing J-lossless conjugator, so does the (TAT^{-1}, TB) for each nonsingular T.

[**5**] Prove that there exists a function $f \in \mathbf{BH}^\infty$ that is real and satisfies (5.47) iff the pair

$$\left(\begin{bmatrix} \sigma & \omega \\ -\omega & \sigma \end{bmatrix}, \begin{bmatrix} 1 & \varepsilon \\ 0 & \delta \end{bmatrix}\right)$$

has a stabilizing J-lossless conjugator, where $\alpha_1 = \sigma + j\omega$ and $\beta_1 = \varepsilon + j\delta$.

[6] Show that for any real β satisfying $-1 < \beta < 1$, there exists a real function $f(s)$ such that $|f(s)| < 1$ for each $\mathrm{Re}[s] > 0$ and

$$f(\sigma + j\omega) = \beta, \quad \sigma > 0, \quad \omega \neq 0.$$

[7] Assume that $BJB^T \leq 0$. Prove that (A, B) has a stabilizing J-lossless conjugator only if A is stable.

[8] Let $f(s) \in \mathbf{BH}_\infty$ be a real function and $f(\alpha + j\beta) = re^{j\theta}$ with $\alpha > 0$, $0 < r < 1$. Show that

$$\frac{\beta^2}{\alpha^2} > \frac{2r^2(1 - \cos 2\theta)}{(1 - r^2)^2}.$$

Chapter 6

J-Lossless Factorizations

6.1 (J, J')-Lossless Factorization and Its Dual

Assume that G is a stable and invertible transfer function. If G^{-1} has a stabilizing J-lossless conjugator Θ, then

$$H := G^{-1}\Theta \tag{6.1}$$

is stable. Due to Lemma 5.3, the zeros of H coincide with those of G^{-1} which are stable from the assumption that G is stable. Hence, H^{-1} is also stable. Writing the relation (6.1) as

$$G = \Theta H^{-1},$$

we see that G is represented as the product of a J-lossless matrix Θ and a unimodular matrix H^{-1}. This is a factorization of G which is of fundamental importance in H^{∞} control theory.

Now we define this factorization in a slightly more general form.

DEFINITION 6.1 *Let G be a rational matrix with $m + r$ rows and $p + r$ columns. If G is represented as the product of a (J_{mr}, J_{pr})-lossless matrix Θ and a unimodular matrix Π (i.e., both Π and Π^{-1} are stable),*

$$G = \Theta \Pi, \tag{6.2}$$

then G is said to have a (J_{mr}, J_{pr})-lossless factorization, or simply (J, J')-lossless factorization when the sizes of signature matrices are clear from the context, or irrelevant.

The factorization (6.2) is a generalization of well-known *inner-outer factorization* which corresponds to the case where J_{mr} is degenerated to the identity matrix ($r = 0$). It should be noted that G in (6.2) can be unstable, whereas the inner-outer factorization is only for stable matrices. Also, the factorization (6.2) includes the well-known spectral factorization of a positive matrix. To see this, let

$$\Gamma(s) = G_1^\sim(s)G_1(s) - G_2^\sim(s)G_2(s),$$

where $G_1 \in \mathbf{H}^\infty_{r\times r}$ and $G_2 \in \mathbf{H}^\infty_{m\times r}$. It is easily seen that

$$\Gamma(j\omega) > 0 \quad \forall\omega \tag{6.3}$$

iff

$$G(s) := \begin{bmatrix} G_2(s) \\ G_1(s) \end{bmatrix}$$

has a (J_{mr}, J_{0r})-lossless factorization, where J_{0r} is identified with $-I_r$. Indeed, if $G(s) = \Theta(s)\Pi(s)$ where $\Theta(s)$ is $(J_{mr}, -I_r)$-lossless and $\Pi(s)$ is unimodular, then $\Theta^\sim(s)J_{mr}\Theta(s) = -I_r$. Hence

$$G^\sim(s)JG(s) = G_2^\sim(s)G_2(s) - G_1^\sim(s)G_1(s) = -\Pi^\sim(s)\Pi(s),$$

which implies (6.3). Moreover, the unimodular factor $\Pi(s)$ gives a spectral factor of a positive matrix $\Gamma(s)$. Thus, the (J, J')-lossless factorization includes the two important factorizations in linear system theory, namely, inner-outer factorization and spectral factorization as special cases.

The factorization (6.2), if it exists, is unique up to the constant J'-unitary matrix. Indeed, if

$$G = \Theta_1\Pi_1 = \Theta_2\Pi_2$$

are two (J, J')-lossless factorizations of G, then we have

$$\Pi_1^\sim J'\Pi_1 = \Pi_2^\sim J'\Pi_2.$$

In other words, $\Pi_1\Pi_2^{-1}$ is J'-unitary. Since $\Pi_1\Pi_2^{-1}$ is unimodular, it is constant; that is, $\Pi_1 = N\Pi_2$ for a constant J'-unitary matrix N. See Problem 4.8. Hence, $\Theta_1 = \Theta_2 N^{-1}$. Thus, the assertion has been established.

Since $\Theta^{\sim}J\Theta = J'$, the relation (6.2) implies

$$G^{\sim}JG = \Pi^{\sim}J'\Pi. \tag{6.4}$$

This representation is usually called a *J-spectral factorization* [36]. From this relation, it follows that $\Theta = G\Pi^{-1}$ is (J,J')-unitary. However, it is not (J,J')-lossless in general. In order to guarantee that $\Theta = G\Pi^{-1}$ is (J,J')-lossless, we need some additional condition on Π.

A dual of (J,J')-lossless factorization is defined for a matrix G with $m+q$ rows and $m+r$ columns. If G is represented as the product of a unimodular matrix Ω and a dual (J,J')-lossless matrix Ψ, that is,

$$G = \Omega\Psi, \tag{6.5}$$

G is said to have a *dual (J,J')-lossless factorization.*

Example 6.1
Let G be given by

$$G = \begin{bmatrix} \dfrac{s-2}{s+2} & -\dfrac{s-2}{s+2} \\ 0 & \dfrac{s+3}{s-1} \end{bmatrix}. \tag{6.6}$$

It is obvious that G has a $J_{1,1}$-lossless factorization (6.2) with factors given by

$$\Theta = \begin{bmatrix} \dfrac{s-2}{s+2} & 0 \\ 0 & \dfrac{s+1}{s-1} \end{bmatrix}, \quad \Pi = \begin{bmatrix} 1 & -1 \\ 0 & \dfrac{s+3}{s+1} \end{bmatrix}. \tag{6.7}$$

It was proven in Example 4.4 that Θ is $J_{1,1}$-lossless. Since

$$\Pi^{-1} = \begin{bmatrix} 1 & \dfrac{s+1}{s+3} \\ 0 & \dfrac{s+1}{s+3} \end{bmatrix},$$

Π is unimodular. Therefore, G in (6.6) has a $J_{1,1}$-lossless factorization with factors given in (6.7).

If $q = r$, Ψ in (6.5) is dual J-lossless. Therefore, the relation (6.5) implies

$$G^{-1} = \Psi^{-1}\Omega^{-1} = J\Psi^{\sim}J\Omega^{-1}. \tag{6.8}$$

Since $\Psi^\sim$ is J-lossless due to Lemma 4.12, $J\Psi^\sim J$ is J-lossless. Therefore, the relation (6.8) represents a J-lossless factorization of G^{-1}. Thus, we have shown the following result:

LEMMA 6.2 *Assume that G is a square invertible matrix. G has a dual J-lossless factorization iff G^{-1} has a J-lossless factorization.*

From the relation (6.5) and $\Psi J \Psi^\sim = J'$, we have

$$GJG^\sim = \Omega J' \Omega^\sim, \tag{6.9}$$

which is dual to (6.4). This relation can be said to be a *dual* (J, J')-*spectral factorization.*

6.2 (J, J')-Lossless Factorization by J-Lossless Conjugation

In this section, it is shown that a (J, J')-lossless factorization is easily carried out by J-lossless conjugation.

We first consider the stable case. Assume that G is stable and has a (J_{mr}, J_{pr})-lossless factorization

$$G = \Theta \Pi,$$

where Θ is (J_{mr}, J_{pr})-lossless and Π is unimodular. Let Θ_1 be a complement of Θ (see the sentence following Lemma 4.7 in Section 4.4.) and let $G_1 := \Theta_1 \Pi_1$ where Π_1 is a unimodular matrix. Now we have

$$\begin{bmatrix} G_1 & G \end{bmatrix} = \begin{bmatrix} \Theta_1 & \Theta \end{bmatrix} \begin{bmatrix} \Pi_1 & 0 \\ 0 & \Pi \end{bmatrix}.$$

Since $\begin{bmatrix} \Theta_1 & \Theta \end{bmatrix}$ is J_{mr}-lossless, $\begin{bmatrix} G_1 & G \end{bmatrix}$ is invertible. This relation is written as

$$\begin{bmatrix} G_1 & G \end{bmatrix}^{-1} \begin{bmatrix} \Theta_1 & \Theta \end{bmatrix} = \begin{bmatrix} \Pi_1^{-1} & 0 \\ 0 & \Pi^{-1} \end{bmatrix}.$$

Since both Π^{-1} and Π_1^{-1} are stable, $\begin{bmatrix} \Theta_1 & \Theta \end{bmatrix}$ must be a J-lossless stabilizing conjugator of $\begin{bmatrix} G_1 & G \end{bmatrix}^{-1}$. Writing

$$\begin{bmatrix} G_1 & G \end{bmatrix}^{-1} = \begin{bmatrix} G^\perp \\ G^\dagger \end{bmatrix}, \tag{6.10}$$

we have $G^\perp \Theta = 0$. We now have the following result.

THEOREM 6.3 *Let $G(s)$ be a stable proper rational $(m+r)\times(p+r)$ matrix with $m \geq p$. It has a (J_{mr}, J_{pr})-lossless factorization iff there exists a G_1 such that*

$$\hat{G} := \begin{bmatrix} G_1 & G \end{bmatrix}$$

is square and invertible and

$$\hat{G}^{-1} = \begin{bmatrix} G^{\perp} \\ G^{\dagger} \end{bmatrix}$$

has a stabilizing J_{mr}-lossless conjugator Θ_0 for which

$$G^{\perp}\Theta_0 D_c = 0 \tag{6.11}$$

for some constant (J_{mr}, J_{pr})-unitary matrix D_c. In that case, factors in (6.2) are given by

$$\Theta = \Theta_0 D_c, \quad \Pi = (G^{\dagger}\Theta)^{-1}. \tag{6.12}$$

Proof. Necessity has already been proven. To show the sufficiency, we write

$$\begin{bmatrix} G_1 & G \end{bmatrix}^{-1}\Theta = \begin{bmatrix} G^{\perp} \\ G^{\dagger} \end{bmatrix}\Theta_0 D_c = \begin{bmatrix} 0 \\ \Pi^{-1} \end{bmatrix}. \tag{6.13}$$

Since Θ_0 is a stabilizing conjugator of $\begin{bmatrix} G_1 & G \end{bmatrix}^{-1}$, Π^{-1} is stable. Since G is stable from the assumption, $G^{\dagger}$ has no unstable zeros. Since the J-lossless conjugation does not create new zeros due to Lemma 5.3, the zeros of $G^{\dagger}\Theta$ are the zeros of $G^{\dagger}$. Hence Π is stable. Thus, Π given in (6.12) is unimodular. From (6.11), $G\Pi^{-1} = GG^{\dagger}\Theta = (I - G_1G^{\perp})\Theta = \Theta$. We see that Θ is a (J, J')-lossless factor of G. ∎

Now we consider the (J, J')-lossless factorization for general unstable systems. It can be shown that the general case can be reduced to the stable case.

Assume that G has a factorization (6.2). According to Lemma 4.9, Θ is represented as a product $\Theta = \Theta_+\Theta_-$ where Θ_+ is an anti-stable J_{mr}-lossless matrix and Θ_- is a stable (J_{mr}, J_{pr})-lossless matrix. From the relation

$$G = \Theta_+\Theta_-\Pi, \tag{6.14}$$

and $\Theta_+^{\sim}J\Theta_+ = J$, we have

$$G^{\sim}J\Theta_+ = \Pi^{\sim}\Theta_-^{\sim}J. \tag{6.15}$$

Since the right-hand side of (6.15) is anti-stable, Θ_+ must be an anti-stabilizing J-lossless conjugator of $G^\sim J$. From (6.14), it follows that

$$J\Theta_+^\sim JG = \Theta_-\Pi. \tag{6.16}$$

This implies that $J\Theta_+^\sim JG$ which is stable has a (J, J')-lossless factorization. The above reasoning is summed up as follows.

THEOREM 6.4 *A matrix with $(m+r)$ rows and $(p+r)$ columns has a (J_{mr}, J_{pr})-lossless factorization iff there exists an anti-stabilizing J_{mr}-lossless conjugator Θ_+ of $G^\sim J_{mr}$ such that a stable matrix $J\Theta_+^\sim J_{mr}G$ has a (J_{mr}, J_{pr})-lossless factorization. In that case, the factors are given by $\Theta = \Theta_+\Theta_-$ and Π where Θ_- and Π are the factors of the factorization for $J_{mr}\Theta_+^\sim J_{mr}G$ given in (6.16).*

Proof. The necessity has already been proven in the preceding argument. The sufficiency is obvious from (6.16) and the identity $\Theta_+ J\Theta_+^\sim = J$. ∎

Theorem 6.4 enables us to describe a procedure of (J, J')-lossless factorization as the following steps.

Step 1. Find an anti-stabilizing J-lossless conjugator $\Theta_+(s)$ of $G^\sim(s)J$.
Step 2. Obtain a (J, J')-lossless factorization of a stable system

$$J(G^\sim J\Theta_+)^\sim = J\Theta_+^\sim JG = \Theta_-\Pi$$

as in (6.16). The factors are given by $\Theta = \Theta_+\Theta_-$ and Π.

Since all the zeros and poles of Π are stable in the representation (6.2), all the unstable zeros and the poles of G must be absorbed in Θ. If λ is an unstable pole of $G(s)$, $-\lambda$ is a stable pole of $G^\sim(s)J$ that is to be conjugated by $\Theta_+(s)$. Due to Lemma 5.3, $\Theta_+(s)$ has λ as its pole. Thus, Step 1 is the process of embedding unstable poles of $G(s)$ in $\Theta(s)$ as the poles of its anti-stable portion $\Theta_+(s)$. In Step 2, since $J\Theta_+^\sim JG$ is stable and its unstable zeros are those of G, its (J, J')-lossless factorization represents the procedure of embedding the unstable zeros of $G(s)$ in $\Theta(s)$ as the zeros of its stable portion $\Theta_-(s)$. In this respect, we call Step 1 a *pole extraction*, and Step 2 a *zero extraction*.

Example 6.2
Consider the system G in Example 6.1. Step 1 is to find an anti-

stabilizing J-lossless conjugator of

$$G^{\sim}(s)J = \begin{bmatrix} \frac{s+2}{s-2} & 0 \\ -\frac{s+2}{s-2} & -\frac{s-3}{s+1} \end{bmatrix}.$$

Since $s = -1$ is the only stable pole of $G^{\sim}(s)J$, we have

$$\Theta_+(s) = \begin{bmatrix} 1 & 0 \\ 0 & \frac{s+1}{s-1} \end{bmatrix}$$

as its anti-stabilizing J-lossless conjugator.
Step 2 is to find a J-lossless factorization for a stable system

$$J\Theta_+^{\sim}(s)JG(s) = \begin{bmatrix} \frac{s-2}{s+2} & -\frac{s-2}{s+2} \\ 0 & \frac{s+3}{s+1} \end{bmatrix}.$$

It is obvious that the factors

$$\Theta_-(s) = \begin{bmatrix} \frac{s-2}{s+2} & 0 \\ 0 & 1 \end{bmatrix}, \quad \Pi(s) = \begin{bmatrix} 1 & -1 \\ 0 & \frac{s+3}{s+1} \end{bmatrix}$$

give a factorization (6.16). Obviously,

$$\begin{aligned} \Theta(s) &= \Theta_+(s)\Theta_-(s) \\ &= \begin{bmatrix} 1 & 0 \\ 0 & \frac{s+1}{s-1} \end{bmatrix} \begin{bmatrix} \frac{s-2}{s+2} & 0 \\ 0 & 1 \end{bmatrix} \end{aligned}$$

gives a factorization of $\Theta(s)$ into the part extracting poles and that extracting zeros.

6.3 (J, J')-Lossless Factorization in State Space

In this section, the state-space theory of (J, J')-lossless factorization is developed based on the results obtained in the preceding section.

Let

$$G(s) = \left[\begin{array}{c|c} A & B \\ \hline C & D \end{array}\right] \tag{6.17}$$

be a state-space realization of a transfer function $G(s)$. The numbers of inputs and outputs are $m+r$ and $p+r$, respectively, which are represented through the size of the D-matrix in (6.17); that is,

$$D \in \mathbf{R}^{(m+r)\times(p+r)}.$$

If $G(s)$ has a factorization (6.2), $G(s)$ is left invertible. Hence, D is of full column rank. Letting $s \to \infty$ in (6.4) yields

$$D^T J D = E^T J' E, \tag{6.18}$$

where $E = \Pi(\infty)$ is nonsingular and we write

$$J = J_{mr} = \begin{bmatrix} I_m & 0 \\ 0 & -I_r \end{bmatrix}, \quad J' = J_{pr} = \begin{bmatrix} I_p & 0 \\ 0 & -I_r \end{bmatrix}.$$

Hence, the existence of E satisfying (6.18) is a necessary condition for (J, J')-lossless factorization.

We first consider the stable case. Based on Theorem 6.3, we can derive a state-space characterization of (J, J')-lossless factorization.

THEOREM 6.5 *Assume that $G(s)$ is stable and its minimal realization is given by (6.17) with $D \in \mathbf{R}^{(m+r)\times(p+r)}$. $G(s)$ has a (J, J')-lossless factorization iff there exist a nonsingular matrix E satisfying (6.18) and a solution $X \geq 0$ of an algebraic Riccati equation*

$$\begin{aligned} XA + A^T X - (C^T JD + XB)(D^T JD)^{-1}(D^T JC + B^T X) \\ + C^T JC = 0 \end{aligned} \tag{6.19}$$

such that

$$\hat{A} := A + BF \tag{6.20}$$

is stable, where

$$F := -(D^T JD)^{-1}(D^T JC + B^T X). \tag{6.21}$$

In that case, the factors (6.2) are given, respectively, by

$$\Theta(s) = \left[\begin{array}{c|c} A+BF & B \\ \hline C+DF & D \end{array}\right] E^{-1}, \quad \Pi(s) = E \left[\begin{array}{c|c} A & -B \\ \hline F & I \end{array}\right], \tag{6.22}$$

where E is a nonsingular matrix satisfying (6.18).

Proof. The sufficiency is almost obvious by observing that the identity (6.2) holds for the factors (6.22) and

$$\Pi(s)^{-1} = \left[\begin{array}{c|c} A+BF & B \\ \hline F & I \end{array}\right] E^{-1}$$

is stable.

The proof of necessity is carried out following the scenario due to Theorem 6.3. First, we augment G as

$$\hat{G} = \left[\begin{array}{cc} G_1 & G \end{array}\right] = \left[\begin{array}{c|cc} A & B_1 & B \\ \hline C & D_1 & D \end{array}\right], \tag{6.23}$$

where D_1 is chosen such that $\hat{D} := \left[\begin{array}{cc} D_1 & D \end{array}\right]$ is invertible. Let

$$\hat{B} := \left[\begin{array}{cc} B_1 & B \end{array}\right], \quad \left[\begin{array}{c} D^{\perp} \\ D^{\dagger} \end{array}\right] := \hat{D}^{-1}.$$

It is clear that $D^{\perp}D = 0$, $D^{\dagger}D = I$. From the inversion law (2.15), it follows that

$$\hat{G}^{-1} = \left[\begin{array}{c} G^{\perp} \\ G^{\dagger} \end{array}\right] = \left[\begin{array}{c|c} A - \hat{B}\hat{D}^{-1}C & \hat{B}\hat{D}^{-1} \\ \hline -\hat{D}^{-1}C & \hat{D}^{-1} \end{array}\right].$$

We compute the J-lossless conjugator Θ_0 of $\hat{G}^{-1}$. Due to Theorem 5.2, $\hat{G}^{-1}$ has a J-lossless conjugator iff there exists a solution $X \geq 0$ of a Riccati equation

$$X(A - \hat{B}\hat{D}^{-1}C) + (A - \hat{B}\hat{D}^{-1}C)^T X - X\hat{B}\hat{D}^{-1}J\hat{D}^{-T}\hat{B}^T X = 0, \tag{6.24}$$

such that

$$\hat{A} := A - \hat{B}\hat{D}^{-1}(C + J\hat{D}^{-T}\hat{B}^T X) \tag{6.25}$$

is stable. In that case, the J-lossless conjugator and the conjugated system are given, respectively, by

$$\Theta_0(s) = \left[\begin{array}{c|c} \hat{A} & \hat{B}\hat{D}^{-1} \\ \hline -J\hat{D}^{-T}\hat{B}^T X & I \end{array}\right],$$

$$\left[\begin{array}{c} G^{\perp}(s) \\ G^{\dagger}(s) \end{array}\right] \Theta_0(s) = \left[\begin{array}{c|c} \hat{A} & \hat{B}\hat{D}^{-1} \\ \hline -\hat{D}^{-1}(C + J\hat{D}^{-T}\hat{B}^T X) & \hat{D} \end{array}\right].$$

Now we find an augmentation (B_1, D_1) in (6.23) and a constant (J, J')-unitary matrix D_c such that

$$G^{\perp}(s)\Theta_0(s)D_c = \left[\begin{array}{c|c} \hat{A} & \hat{B}\hat{D}^{-1} \\ \hline -D^{\perp}(C + J\hat{D}^{-T}\hat{B}^T X) & D^{\perp} \end{array}\right] D_c = 0. \tag{6.26}$$

From $D^{\perp}D_c = 0$ and D_c is (J, J')-unitary, we take $D_c = DE^{-1}$, where E satisfies (6.18). Hence $\hat{B}\hat{D}^{-1}D_c = (B_1 D^{\perp} + BD^{\dagger})DE^{-1} = BE^{-1}$. Since (A, B) is controllable from the assumption, $(\hat{A}, BE^{-1})$ is controllable. The condition (6.26) holds iff

$$D^{\perp}(C + J\hat{D}^{-T}\hat{B}^T X) = 0. \tag{6.27}$$

The problem is to find an augmentation satisfying (6.27). The relation (6.27) implies that

$$C + J\hat{D}^{-T}\hat{B}^T X = -DF \tag{6.28}$$

for some F. Premultiplication of both sides of (6.28) by $D^T J$ verifies that F is given by (6.21).

Now Equation (6.24) is rewritten as

$$XA + A^T X - (C^T + X\hat{B}\hat{D}^{-1}J)J(C + J\hat{D}^{-T}\hat{B}^T X) + C^T JC = 0.$$

Substituting (6.28) in this identity yields

$$XA + A^T X - F^T D^T JDF + C^T JC = 0,$$

which is identical to (6.19). Thus, we have seen that X must satisfy the Riccati equation (6.19) which is independent of the augmentation (B_1, D_1). It is easy to see that $\hat{A}$ in (6.25) is given by (6.20) due to (6.28). ∎

Remark: It is worth noting that any augmentation (B_1, D_1) such that

$$D_1^T J(C + DF) = B_1^T X$$

satisfies (6.27).

Example 6.3

Let us compute a $J_{1,1}$-lossless factorization of

$$G(s) = \begin{bmatrix} \dfrac{s-2}{s+2} & -\dfrac{s-2}{s+2} \\ 0 & \dfrac{s+3}{s+1} \end{bmatrix} = \left[\begin{array}{cc|cc} -2 & 0 & -2 & 2 \\ 0 & -1 & 0 & 1 \\ \hline 2 & 0 & 1 & -1 \\ 0 & 2 & 0 & 1 \end{array}\right].$$

It can be seen that

$$X = \begin{bmatrix} 1 & 0 \\ 0 & 0 \end{bmatrix} \geq 0$$

is a stabilizing solution of (6.19) for this case with

$$\hat{A} = \begin{bmatrix} -2 & 0 \\ 0 & -3 \end{bmatrix}.$$

Hence, due to Theorem 6.5, $G(s)$ has a $J_{1,1}$-lossless factorization. The factors in (6.22) are calculated to be

$$\Theta(s) = \left[\begin{array}{c|cc} -2 & -2 & 0 \\ \hline 2 & 1 & 0 \\ 0 & 0 & 1 \end{array}\right] = \begin{bmatrix} \frac{s-2}{s+2} & 0 \\ 0 & 1 \end{bmatrix},$$

$$\Pi(s) = \left[\begin{array}{c|cc} -1 & 0 & 1 \\ \hline 0 & 1 & -1 \\ 2 & 0 & 1 \end{array}\right] = \begin{bmatrix} 1 & -1 \\ 0 & \frac{s+3}{s+1} \end{bmatrix}.$$

Now we move on to the general case where G is unstable. In this case, the factorization is more complicated and we have to solve two Riccati equations, instead of one for stable cases.

THEOREM 6.6 *Let (6.17) be a minimal realization of G with $D \in \mathbf{R}^{(m+r)\times(p+r)}$. It has a (J_{mr}, J_{pr})-lossless factorization iff*

(i) there exists a nonsingular matrix E satisfying (6.18),

(ii) there exists a solution $X \geq 0$ of the algebraic Riccati equation (6.19) such that $\hat{A}$ in (6.20) is stable,

(iii) there exists a solution $\bar{X} \geq 0$ of the algebraic Riccati equation

$$\bar{X}A^T + A\bar{X} + \bar{X}C^T JC\bar{X} = 0 \tag{6.29}$$

such that

$$\bar{A} := A + \bar{X}C^T JC \tag{6.30}$$

is stable, and

(iv)

$$\sigma(X\bar{X}) < 1. \tag{6.31}$$

In that case, the factors are given, respectively, by

$$\Theta(s) = \left[\begin{array}{cc|c} -\bar{A}^T & 0 & \begin{bmatrix} I & -X \\ -\bar{X} & I \end{bmatrix}^{-1} \begin{bmatrix} C^T JD \\ B \end{bmatrix} \\ 0 & A+BF & \\ \hline -C\bar{X} & C+DF & D \end{array}\right] E^{-1} \tag{6.32}$$

$$\Pi(s) = E \left[\begin{array}{c|c} A+\bar{X}C^T JC & -(B+\bar{X}C^T JD) \\ \hline F(I-\bar{X}X)^{-1} & I \end{array}\right], \tag{6.33}$$

where E is a matrix satisfying (6.18), and F is given by (6.21).

Proof. The proof follows faithfully the procedure described in the preceding section.

The first step is to obtain an anti-stabilizing J-lossless conjugator $\Theta_+(s)$ of a system $G^\sim(s)J$ whose state-space realization is given by

$$G^\sim(s)J = \left[\begin{array}{c|c} -A^T & C^T J \\ \hline -B^T & D^T J \end{array}\right]. \tag{6.34}$$

Due to Theorem 5.2, an anti-stabilizing J-lossless conjugator of (6.34) exists iff there exists a solution $\bar{X} \geq 0$ of the algebraic Riccati equation

$$-\bar{X}A^T - A\bar{X} - \bar{X}C^T JC\bar{X} = 0$$

such that $-A^T - C^T JC\bar{X}$ is anti-stable. This proves the necessity of (iii).

According to (5.5), a desired anti-stabilizing J-lossless conjugator Θ_+ is given by

$$\Theta_+(s) = \left[\begin{array}{c|c} -\bar{A}^T & C^T J \\ \hline -C\bar{X} & I \end{array}\right]. \tag{6.35}$$

The conjugated system is given by

$$G^\sim(s)J\Theta_+(s) = \left[\begin{array}{c|c} -\bar{A}^T & -C^T J \\ \hline B^T + D^T JC\bar{X} & D^T J \end{array}\right].$$

Hence, we have

$$J\Theta_+^{\sim}(s)JG(s) = \left[\begin{array}{c|c} \bar{A} & B + \bar{X}C^TJD \\ \hline C & D \end{array}\right]. \tag{6.36}$$

Now we compute a (J, J')-lossless factorization of (6.36) based on Theorem 6.5. A (J, J')-lossless factorization of (6.36) exists iff there exists a solution $Z \geq 0$ of the algebraic Riccati equation

$$\begin{aligned} Z\bar{A} + \bar{A}^TZ - (C^TJD + Z(B + \bar{X}C^TJD))(D^TJD)^{-1} \\ (D^TJC + (B + \bar{X}C^TJD)^TZ) + C^TJC = 0 \end{aligned} \tag{6.37}$$

such that $\bar{A} + (B + \bar{X}C^TJD)\hat{F}$ is stable where

$$\hat{F} := -(D^TJD)^{-1}(D^TJC + (B + \bar{X}C^TJD)^TZ) \tag{6.38}$$

is stable. From (6.29) and (6.30), it follows that

$$\begin{aligned} &Z\bar{A} + \bar{A}^TZ + C^TJC = \\ &\quad ZA(I + \bar{X}Z) + (I + Z\bar{X})A^TZ + (I + Z\bar{X})C^TJC(I + \bar{X}Z). \end{aligned}$$

Therefore, premultiplication by $(I + Z\bar{X})^{-1}$ and postmultiplication by $(I + \bar{X}Z)^{-1}$ of (6.37) using this relation yield

$$\begin{aligned} XA + A^TX - (C^TJD + XB)(D^TJD)^{-1}(D^TJC + B^TX) \\ + C^TJC = 0 \end{aligned} \tag{6.39}$$

where we take $X = (I + Z\bar{X})^{-1}Z$, or equivalently,

$$Z = X(I - \bar{X}X)^{-1} = (I - X\bar{X})^{-1}X. \tag{6.40}$$

Now, from (6.38) and (6.40), it follows that

$$\begin{aligned} \hat{F} &= -(D^TJD)^{-1}(D^TJC(I + \bar{X}Z) + B^TZ) \\ &= F(I - \bar{X}X)^{-1}, \end{aligned}$$

where F is given by (6.21). Also, from (6.29), (6.30), and (6.39), it follows that

$$\begin{aligned}\bar{A}(I-\bar{X}X) &= A+\bar{X}(C^TJC+A^TX)\\ &= A-\bar{X}((C^TJD+XB)F+XA)\\ &= (I-\bar{X}X)(A+BF)-(B+\bar{X}C^TJD)F. \end{aligned} \tag{6.41}$$

Therefore, we have

$$(\bar{A}+(B+\bar{X}C^TJD)\hat{F})(I-\bar{X}X)=(I-\bar{X}X)(A+BF).$$

Since $\bar{A}+(B+\bar{X}C^TJD)\hat{F}$ is stable, so is

$$A+BF=(I-\bar{X}X)^{-1}(\bar{A}+(B+\bar{X}C^TJD)\hat{F})(I-\bar{X}X).$$

Thus, we have established the necessity of (ii).

The necessity of the condition (iv) follows from (6.40) and $Z\geq 0$.

To prove the sufficiency, we can check the validity of the representation (6.2) by direct computations. The following two lemmas, which are interesting in their own lights, give the proof of the sufficiency.

LEMMA 6.7 *Let $\Theta(s)$ in (6.32) be represented as*

$$\Theta(s)=\left[\begin{array}{c|c} A_\theta & B_\theta\\ \hline C_\theta & D_\theta\end{array}\right],$$

that is,

$$\begin{aligned}A_\theta &:= \begin{bmatrix}-(A+\bar{X}C^TJC)^T & 0\\ 0 & A+BF\end{bmatrix},\\ B_\theta &:= \begin{bmatrix} I & -X\\ -\bar{X} & I\end{bmatrix}^{-1}\begin{bmatrix}C^TJD_\theta\\ BE^{-1}\end{bmatrix},\\ C_\theta &:= [-C\bar{X}\quad C+DF],\quad D_\theta:=DE^{-1}.\end{aligned}$$

If $\bar{X}$ and X satisfying (ii) $\sim$ (iv) in Theorem 6.5 exist, then

$$P:=\begin{bmatrix}\bar{X} & 0\\ 0 & X\end{bmatrix}\begin{bmatrix} I & -X\\ -\bar{X} & I\end{bmatrix}=\begin{bmatrix} I & -\bar{X}\\ -X & I\end{bmatrix}\begin{bmatrix}\bar{X} & 0\\ 0 & X\end{bmatrix} \tag{6.42}$$

satisfies the equations

$$PA_\theta+A_\theta^TP+C_\theta^TJC_\theta=0,\quad B_\theta^TP+D_\theta^TJC_\theta=0. \tag{6.43}$$

Proof. The Riccati equations (6.19) and (6.29) can be written, respectively, as

$$X(A+BF)+(A+BF)^TX+(C+DF)^TJ(C+DF)=0,$$
$$\bar{X}\bar{A}^T+\bar{A}\bar{X}-\bar{X}C^TJC\bar{X}=0,$$

where F is given by (6.21). Using the preceding equations, we can show the identities (6.43). ∎

The preceding lemma implies that Θ given in (6.32) is (J, J')-lossless according to Theorem 4.5. We have used the identity (6.41), which can be written as

$$\begin{aligned}\bar{A}+(B+\bar{X}C^TJD)F(I-\bar{X}X)^{-1}\\ =(I-\bar{X}X)(A+BF)(I-\bar{X}X)^{-1}.\end{aligned} \qquad (6.44)$$

This relation implies the following lemma which is proven by direct manipulations.

LEMMA 6.8 $\Pi(s)$ *given by (6.33) has the inverse*

$$\Pi^{-1}=\left[\begin{array}{c|c} A+BF & (I-\bar{X}X)^{-1}(B+\bar{X}C^TJD) \\ \hline F & I \end{array}\right]E^{-1}. \qquad (6.45)$$

This lemma shows that Π is unimodular because $A+BF$ is stable.

(*Proof of the Sufficiency of Theorem 6.6*)

Now we check that the relation (6.2) actually holds. From (6.17), (6.45), and the concatenation rule (2.13), we have

$$G\Pi^{-1}=\left[\begin{array}{cc|c} A & BF & B \\ 0 & A+BF & (I-\bar{X}X)^{-1}(B+\bar{X}C^TJD) \\ \hline C & DF & D \end{array}\right]E^{-1}$$
$$=\left[\begin{array}{cc|c} A & 0 & -\bar{X}(I-X\bar{X})^{-1}(XB+C^TJD) \\ 0 & A+BF & (I-\bar{X}X)^{-1}(B+\bar{X}C^TJD) \\ \hline C & C+DF & D \end{array}\right]E^{-1}.$$

Since $A\bar{X} = -\bar{A}^T\bar{X}$ from (6.29), we have

$$C(sI - A)^{-1}\bar{X} = C\bar{X}(sI + \bar{A}^T)^{-1}.$$

Also, from the identity

$$\begin{bmatrix} I & -X \\ -\bar{X} & I \end{bmatrix}^{-1} = \begin{bmatrix} (I - X\bar{X})^{-1} & 0 \\ 0 & (I - \bar{X}X)^{-1} \end{bmatrix} \begin{bmatrix} I & X \\ \bar{X} & I \end{bmatrix},$$

we see that $G\Pi^{-1}$ is identical to Θ as given in (6.32). ∎

Example 6.4
Consider again the system G in Example 6.1 whose state-space realization is given by

$$G(s) = \left[\begin{array}{cc|cc} -2 & 0 & -2 & 2 \\ 0 & 1 & 0 & 2 \\ \hline 2 & 0 & 1 & -1 \\ 0 & 2 & 0 & 1 \end{array}\right] = \left[\begin{array}{c|c} A & B \\ \hline C & D \end{array}\right].$$

Noting that $C^TJD = -B$ in this case, we easily see that

$$X = \begin{bmatrix} 1 & 0 \\ 0 & 0 \end{bmatrix}$$

is the stabilizing solution to the Riccati equation (6.19). Also, it is easy to see that

$$\bar{X} = \begin{bmatrix} 0 & 0 \\ 0 & 1/2 \end{bmatrix}$$

is the stabilizing solution to the Riccati equation (6.29). The inequality (6.31) is obviously satisfied. We have

$$\hat{A} = A + BF = \begin{bmatrix} -2 & 0 \\ 0 & -3 \end{bmatrix},$$

$$\bar{A} = A + \bar{X}C^TJC = \begin{bmatrix} -2 & 0 \\ 0 & -1 \end{bmatrix}.$$

The J-lossless factor (6.32) is calculated to be

$$\Theta(s) = \left[\begin{array}{c|c} \begin{matrix} 2 & 0 & 0 & 0 \\ 0 & 1 & 0 & 0 \\ 0 & 0 & -2 & 0 \\ 0 & 0 & 0 & -3 \end{matrix} & \begin{bmatrix} 1 & 0 & 1 & 0 \\ 0 & 1 & 0 & 0 \\ 0 & 0 & 1 & 0 \\ 0 & 1/2 & 0 & 1 \end{bmatrix} \begin{bmatrix} 2 & -2 \\ 0 & -2 \\ -2 & 2 \\ 0 & 2 \end{bmatrix} \\ \hline \begin{matrix} 0 & 0 & 2 & 0 \\ 0 & -1 & 0 & 0 \end{matrix} & \begin{bmatrix} 1 & -1 \\ 0 & 1 \end{bmatrix} \end{array}\right] \begin{bmatrix} 1 & 1 \\ 0 & 1 \end{bmatrix},$$

where we took $E = D$. After eliminating the uncontrollable and unobservable portions, we get

$$\Theta(s) = \left[\begin{array}{cc|cc} 1 & 0 & 0 & -2 \\ 0 & -2 & -2 & 0 \\ \hline 0 & 2 & 1 & 0 \\ -1 & 0 & 0 & 1 \end{array}\right]$$

which coincides with the one obtained in (6.7). Also, simple manipulation shows that $\Pi(s)$ in (6.33) is given by

$$\Pi(s) = \left[\begin{array}{c|cc} -1 & 0 & 1 \\ \hline 0 & 1 & -1 \\ 2 & 0 & 1 \end{array}\right],$$

which again coincides with the one given in (6.7).

Example 6.5

Consider a tall system given by

$$G(s) = \left[\begin{array}{cc|cc} 3 & 0 & -\sqrt{2} & 4-\sqrt{2} \\ 0 & -3 & \sqrt{2}/2 & \sqrt{2}/2 \\ \hline 0 & -4 & \sqrt{2}/2 & \sqrt{2}/2 \\ 1 & -2 & \sqrt{2}/2 & \sqrt{2}/2 \\ 2 & -1 & 0 & 1 \end{array}\right].$$

We compute a (J_{21}, J_{11})-lossless factorization based on Theorem 6.6.

A matrix E satisfying (6.18) is given by

$$E = \begin{bmatrix} 1 & 1 \\ 0 & 1 \end{bmatrix}.$$

The stabilizing solution to (6.19) is given by

$$X = \begin{bmatrix} 1/14 & 0 \\ 0 & 2 \end{bmatrix}$$

which is positive definite. Hence, Condition (ii) of Theorem 6.6 is satisfied. For this solution,

$$\hat{A} = A + BF = \begin{bmatrix} -3 & 0 \\ -3/7 & -1 \end{bmatrix},$$

$$F = \begin{bmatrix} (12 - 3\sqrt{2})/7 & 2\sqrt{2} - 1 \\ -12/7 & 1 \end{bmatrix}.$$

In the same way, we can compute the stabilizing solution $\bar{X}$ of (6.29) as

$$\bar{X} = \begin{bmatrix} 2 & 0 \\ 0 & 0 \end{bmatrix} \geq 0$$

for which

$$\bar{A} = \begin{bmatrix} -3 & 0 \\ 0 & -3 \end{bmatrix}.$$

Hence, Condition (iii) of Theorem 6.6 holds. Since

$$X\bar{X} = \begin{bmatrix} 1/7 & 0 \\ 0 & 0 \end{bmatrix},$$

Condition (iv) also holds. Therefore, $G(s)$ has a (J_{21}, J_{11})-lossless factorization. The factors are calculated to be

$$\Theta(s) = \left[\begin{array}{cc|cc} -1 & 0 & -\sqrt{2}/2 & 0 \\ 0 & 3 & -\sqrt{2} & 4 \\ \hline 2 & 0 & \sqrt{2}/2 & 0 \\ 0 & 1 & \sqrt{2}/2 & 0 \\ 0 & 2 & 0 & 1 \end{array}\right],$$

$$\Pi(s) = \left[\begin{array}{c|cc} -3 & \sqrt{2}/2 & \sqrt{2}/2 \\ \hline -2\sqrt{2} & 1 & 1 \\ -1 & 0 & 1 \end{array}\right].$$

6.4 Dual (J, J')-Lossless Factorization in State Space

The complete dualization of the preceding section leads to the derivation of a dual (J, J')-lossless factorization (6.5). In this section, we identify

$$J = J_{mr}, \quad J' = J_{mq}. \tag{6.46}$$

The augmentation required in this section is given by

$$\hat{G} = \begin{bmatrix} G \\ G_2 \end{bmatrix}, \tag{6.47}$$

instead of (6.10), such that

$$\hat{G}^{-1} = \begin{bmatrix} G^{\dagger} & G^{\perp} \end{bmatrix}$$

exists. A condition for the dual (J, J')-lossless factorization of stable systems is given that corresponds to Theorem 6.3.

THEOREM 6.9 *Let $G(s)$ be a stable proper rational $(m+q) \times (m+r)$ matrix. It has a dual (J_{mr}, J_{mq})-lossless factorization iff there exists a G_2 such that*

$$\hat{G} = \begin{bmatrix} G \\ G_2 \end{bmatrix}$$

is square and invertible and $(\hat{G}^{-1})^{\sim}$ has an anti-stabilizing J_{mr}-lossless conjugator Θ_0 for which

$$(G^{\perp})^{\sim}\Theta_0 D_c^T = 0, \quad D_c J_{mr} D_c^T = J_{mq} \tag{6.48}$$

for some constant matrix D_c. In that case, factors in (6.5) are given by

$$\Psi = D_c \Theta_0^{\sim}, \quad \Omega = (\Psi G^{\dagger})^{-1}, \tag{6.49}$$

where $\hat{G}^{-1} = \begin{bmatrix} G^{\dagger} & G^{\perp} \end{bmatrix}$.

Based on the Theorems 6.9 and 5.2, we can show the state-space characterization of dual (J, J')-lossless factorization in terms of its realization (6.17). First, we note that the relation (6.9) implies that

$$DJD^T = EJ'E^T, \tag{6.50}$$

where $E = \Omega(\infty)$. Note that E is nonsingular since $\Omega(s)$ is unimodular.

THEOREM 6.10 *Assume that $G(s)$ is stable and its minimal realization is given by (6.17) with $D \in \mathbf{R}^{(m+q)\times(m+r)}$. $G(s)$ has a dual (J, J')-lossless factorization iff there exists a nonsingular matrix E satisfying (6.50) and a solution $Y \geq 0$ of an algebraic Riccati equation*

$$\begin{aligned} YA^T + AY + (BJD^T - YC^T)(DJD^T)^{-1}(DJB^T - CY) \\ -BJB^T = 0 \end{aligned} \tag{6.51}$$

such that

$$\hat{A} := A + LC \tag{6.52}$$

is stable where

$$L := -(BJD^T - YC^T)(DJD^T)^{-1}. \tag{6.53}$$

In that case, factors in (6.5) are given, respectively, by

$$\Omega = \left[\begin{array}{c|c} A & L \\ \hline -C & I \end{array}\right] E, \quad \Psi = E^{-1}\left[\begin{array}{c|c} \hat{A} & B + LD \\ \hline C & D \end{array}\right], \tag{6.54}$$

where E is a nonsingular matrix satisfying (6.50).

Proof. An augmentation (6.47) is represented in the state space as

$$\hat{G} = \begin{bmatrix} G \\ G_2 \end{bmatrix} = \left[\begin{array}{c|c} A & B \\ \hline C & D \\ C_2 & D_2 \end{array}\right] = \left[\begin{array}{c|c} A & B \\ \hline \hat{C} & \hat{D} \end{array}\right].$$

Due to inversion rule (2.15), we have

$$\hat{G}^{-1} = \begin{bmatrix} G^\dagger & G^\perp \end{bmatrix} = \left[\begin{array}{c|c} A - B\hat{D}^{-1}\hat{C} & B\hat{D}^{-1} \\ \hline -\hat{D}^{-1}\hat{C} & \hat{D}^{-1} \end{array}\right].$$

Due to Theorem 6.9, a dual (J, J')-lossless factorization of G exists iff

$$(\hat{G}^{-1})^{\sim} = \left[\begin{array}{c|c} -(A - B\hat{D}^{-1}\hat{C})^T & \hat{C}^T\hat{D}^{-T} \\ \hline \hat{D}^{-T}B^T & \hat{D}^{-T} \end{array}\right] \tag{6.55}$$

has an anti-stabilizing J-lossless conjugator Θ_0 such that (6.48) holds for some D_c. Equation (5.3) in this case is given by

$$\begin{aligned} Y(A - B\hat{D}^{-1}\hat{C})^T + (A - B\hat{D}^{-1}\hat{C})Y \\ +Y\hat{C}^T\hat{D}^{-T}J\hat{D}^{-1}\hat{C}Y = 0 \end{aligned} \tag{6.56}$$

and Y must stabilize

$$\hat{A} = A - (B - Y\hat{C}^T\hat{D}^{-T}J)\hat{D}^{-1}\hat{C}. \tag{6.57}$$

From (5.5) and (5.6), it follows that

$$\Theta_0 = \left[\begin{array}{c|c} -\hat{A}^T & \hat{C}^T\hat{D}^{-T} \\ \hline -J\hat{D}^{-1}\hat{C}Y & I \end{array}\right], \tag{6.58}$$

$$(\hat{G}^{-1})^{\sim}\Theta_0 = \left[\begin{array}{c|c} -\hat{A}^T & \hat{C}^T\hat{D}^{-T} \\ \hline \hat{D}^{-T}(B^T - J\hat{D}^{-1}\hat{C}Y) & \hat{D}^{-T} \end{array}\right]. \tag{6.59}$$

Therefore, we have

$$(G^{\perp})^{\sim}\Theta_0 D_c^T = \left[\begin{array}{c|c} -\hat{A}^T & \hat{C}^T\hat{D}^{-T} \\ \hline (D^{\perp})^T(B^T - J\hat{D}^{-1}\hat{C}Y) & (D^{\perp})^T \end{array}\right] D_c^T,$$

where we write $\hat{D}^{-1} := \begin{bmatrix} D^{\dagger} & D^{\perp} \end{bmatrix}$. In order that (6.48) holds, $(D^{\perp})^T D_c = 0$. If we take $D_c = (E^{-1}D)^T$ where E satisfies (6.50), this condition is satisfied. In that case, $\hat{C}^T\hat{D}^{-T}D_c = C^T E^{-T}$. Since $(\hat{A}, C)$ is observable, the first relation of (6.48) holds iff

$$(D^{\perp})^T(B^T - J\hat{D}^{-1}\hat{C}Y) = 0,$$

which implies that

$$B - Y\hat{C}^T\hat{D}^{-T}J = -LD \tag{6.60}$$

for some L. By the postmultiplication of both sides of (6.60) by $JD^T$and using $\hat{C}^T(D\hat{D}^{-1})^T = C^T$, we can show that L is given by (6.53). Rewriting (6.57) as

$$YA^T + AY + (B - Y\hat{C}^T\hat{D}^{-T}J)J(B^T - J\hat{D}^{-1}\hat{C}Y) - BJB^T = 0$$

and using (6.60) yield

$$YA^T + AY + LDJD^TL^T - BJB^T = 0.$$

Therefore, we have obtained (6.51) in view of (6.53).

It is easily seen that $\hat{A}$ in (6.57) is given by (6.52). Also, straightforward manipulations show that $\Psi = D_c\Theta_0^{\sim}$ is given by (6.54) with $D_c^T = D^TE^{-T}$. Since

$$\hat{D}^{-T}(B^T - J\hat{D}^{-1}\hat{C}Y) = -\hat{D}^{-T}D^TL^T = -\begin{bmatrix} L^T \\ 0 \end{bmatrix},$$

we have, from (6.59),

$$\Psi G^{\dagger} = E^{-1}\left[\begin{array}{c|c} \hat{A} & L \\ \hline C & I \end{array}\right].$$

Therefore, Ω in (6.49) is given by (6.54). ∎

Now we move on to general unstable cases. The dual version of the factorization used in (6.14) is given by

$$\Psi = \Psi_-\Psi_+, \tag{6.61}$$

where Ψ_- is a stable dual (J, J')-lossless matrix, and Ψ_+ is an anti-stable dual J-lossless matrix. In exactly the same way as in Theorem 6.4, we can show a condition for the existence of a dual (J, J')-lossless factorization.

THEOREM 6.11 *A matrix G with $(m+q)$ rows and $(m+r)$ columns has a dual (J_{mr}, J_{mq})-lossless factorization iff there exists a stabilizing J_{mr}-lossless conjugator Ψ_+ of GJ such that the stable system $GJ\Psi_+J$ has a dual (J_{mr}, J_{mq})-lossless factorization*

$$GJ\Psi_+J = \Omega\Psi_-.$$

In that case, the factors are given, respectively, by Ω and $\Psi = \Psi_-\Psi_+$.

Based on Theorem 6.11, we can describe a procedure of dual (J, J')-lossless factorization as the following steps.

Step 1 : Find a stabilizing J-lossless conjugator $\Psi_+(s)$ of $G(s)J$.
Step 2 : Obtain a dual (J, J')-lossless factorization of a stable system

$$G(s)J\Psi_+(s)J = \Omega\Psi_-(s).$$

The factors are given by $\Psi = \Psi_-\Psi_+$ and Ω, respectively.

Now a state-space characterization of dual (J, J')-lossless factorization is given. The proof is left to the reader (See Problem 6.7).

THEOREM 6.12 *Let (6.17) be a minimal realization of G with $D \in \mathbf{R}^{(m+q)\times(m+r)}$. It has a dual (J_{mr}, J_{mq})-lossless factorization iff*

(i) *there exists a nonsingular matrix E satisfying (6.50),*

(ii) *there exists a solution $Y \geq 0$ of the algebraic Riccati equation (6.51) such that $\hat{A}$ in (6.52) is stable,*

(iii) *there exists a solution $\bar{Y} \geq 0$ of the algebraic Riccati equation*

$$\bar{Y}A + A^T\bar{Y} - \bar{Y}BJB^T\bar{Y} = 0 \tag{6.62}$$

such that

$$\bar{A} := A - BJB^T\bar{Y}$$

is stable, and

(iv) $\sigma(Y\bar{Y}) < 1$.

In that case, factors in (6.5) are given, respectively, by

$$\Psi = E^{-1}\left[\begin{array}{cc|c} \bar{A} & 0 & B + LD \\ 0 & -\hat{A}^T & \bar{Y}B \\ \hline \begin{bmatrix} C & DJB^T \end{bmatrix} & \begin{bmatrix} I & Y \\ \bar{Y} & I \end{bmatrix}^{-1} & D \end{array}\right], \tag{6.63}$$

$$\Omega = \left[\begin{array}{c|c} \bar{A} & -(I - Y\bar{Y})^{-1}L \\ \hline C - DJB^T\bar{Y} & I \end{array}\right]E. \tag{6.64}$$

Using (6.51) and (6.62), we have

$$\begin{aligned} A - BJB^T\bar{Y} + (I - Y\bar{Y})^{-1}L(C - DJB^T\bar{Y}) \\ = (I - Y\bar{Y})^{-1}(A + LC)(I - Y\bar{Y}), \end{aligned} \tag{6.65}$$

which corresponds to (6.44). From this, it follows that

$$\Omega^{-1} = E^{-1}\left[\begin{array}{c|c} A + LC & L \\ \hline (C - DJB^T\bar{Y})(I - Y\bar{Y})^{-1} & I \end{array}\right]. \tag{6.66}$$

6.5 Hamiltonian Matrices

In the preceding sections, we have derived four Riccati equations associated with the (J,J')-lossless factorization and its dual, namely, (6.19), (6.29) for the (J,J')-lossless factorization and (6.51), (6.62) for the dual (J,J')-lossless factorization. If $G(s)$ has a state-space realization (6.17), then direct manipulations using (2.13) yield

$$G^{\sim}(s)JG(s) = \left[\begin{array}{cc|c} A & 0 & B \\ -C^TJC & -A^T & -C^TJD \\ \hline D^TJC & B^T & D^TJD \end{array}\right] \tag{6.67}$$

$$= \left[\begin{array}{cc|c} -A^T & -C^TJC & -C^TJD \\ 0 & A & B \\ \hline B^T & D^TJC & D^TJD \end{array}\right]. \tag{6.68}$$

Associated with these realizations, write

$$f\left(\left[\begin{array}{c|c} A & B \\ \hline C & D \end{array}\right]\right) := \left[\begin{array}{cc} A & 0 \\ -C^TJC & -A^T \end{array}\right] + \left[\begin{array}{c} -B \\ C^TJD \end{array}\right](D^TJD)^{-1}\left[\begin{array}{cc} D^TJC & B^T \end{array}\right] \tag{6.69}$$

$$g\left(\left[\begin{array}{c|c} A & B \\ \hline C & D \end{array}\right]\right) := \left[\begin{array}{cc} A^T & C^TJC \\ 0 & -A \end{array}\right]. \tag{6.70}$$

The matrix $f\left(\left[\begin{array}{c|c} A & B \\ \hline C & D \end{array}\right]\right)$ is the A-matrix of $(G^{\sim}JG)^{-1}$ associated with the realization (6.67), and $g\left(\left[\begin{array}{c|c} A & B \\ \hline C & D \end{array}\right]\right)$ is the A-matrix of $G^{\sim}JG$ associated with the realization (6.68) with the sign reversed. It is straightforward to see that both matrices are Hamiltonian in the sense that they satisfy (3.33).

The Riccati equations (6.19) and (6.29) are associated with the Hamiltonian matrices (6.69) and (6.70), respectively. Actually, X in Theorem 6.6 is represented as $Ric\left(f\left(\left[\begin{array}{c|c} A & B \\ \hline C & D \end{array}\right]\right)\right)$ and $\bar{X}$ in Theorem 6.5 is represented as $Ric\left(g\left(\left[\begin{array}{c|c} A & B \\ \hline C & D \end{array}\right]\right),\right)$ which are written respectively as

$$Ric\ f\left(\left[\begin{array}{c|c} A & B \\ \hline C & D \end{array}\right]\right), \quad Ric\ g\left(\left[\begin{array}{c|c} A & B \\ \hline C & D \end{array}\right]\right).$$

Therefore, the existence condition for (J,J')-lossless factorization stated in Theorem 6.6 is represented as follows.

THEOREM 6.13 *$G(s)$ given by (6.17) has a (J,J')-lossless factorization iff*

(i) *there exists a nonsingular matrix E satisfying (6.18),*

(ii) $X = Ric\ f\left(\left[\begin{array}{c|c} A & B \\ \hline C & D \end{array}\right]\right) \geq 0$ *and* $\bar{X} = Ric\ g\left(\left[\begin{array}{c|c} A & B \\ \hline C & D \end{array}\right]\right) \geq 0$ *exist such that* $\sigma(X\bar{X}) < 1$.

The duals of the above Hamiltonian matrices introduced in (6.69) and (6.70) are defined as

$$f^{\sim}\left(\left[\begin{array}{c|c} A & B \\ \hline C & D \end{array}\right]\right) := -f\left(\left[\begin{array}{c|c} A & B \\ \hline C & D \end{array}\right]^{\sim}\right) = -f\left(\left[\begin{array}{c|c} -A^T & C^T \\ \hline -B^T & D^T \end{array}\right]\right)$$

$$= \begin{bmatrix} A^T & 0 \\ BJB^T & -A \end{bmatrix} - \begin{bmatrix} C^T \\ BJD^T \end{bmatrix} (DJD^T)^{-1} \begin{bmatrix} DJB^T & -C \end{bmatrix} \tag{6.71}$$

and

$$g^{\sim}\left(\left[\begin{array}{c|c} A & B \\ \hline C & D \end{array}\right]\right) := -g\left(\left[\begin{array}{c|c} A & B \\ \hline C & D \end{array}\right]^{\sim}\right) = -g\left(\left[\begin{array}{c|c} -A^T & C^T \\ \hline -B^T & D^T \end{array}\right]\right)$$

$$= \begin{bmatrix} A & -BJB^T \\ 0 & -A^T \end{bmatrix}. \tag{6.72}$$

These are associated with the state-space forms

$$G(s)JG^{\sim}(s) = \left[\begin{array}{cc|c} -A^T & 0 & C^T \\ -BJB^T & A & BJD^T \\ \hline -DJB^T & C & DJD^T \end{array}\right] \tag{6.73}$$

$$= \left[\begin{array}{cc|c} A & -BJB^T & BJD^T \\ 0 & -A^T & C^T \\ \hline C & -DJB^T & DJD^T \end{array}\right]. \tag{6.74}$$

The Riccati equations (6.51) and (6.62) are associated with the Hamiltonian matrices (6.71) and (6.72). Thus, in view of Theorem 6.12, we have the following existence condition for the dual (J,J')-lossless factorization.

THEOREM 6.14 *There exists a dual (J,J')-lossless factorization for $G(s)$ in (6.17) iff*

(i) *there exists a nonsingular E satisfying (6.50), and*

(ii) $Y = Ric\ f^{\sim}\left(\left[\begin{array}{c|c} A & B \\ \hline C & D \end{array}\right]\right) \geq 0$ *and* $\bar{Y} = Ric\ g^{\sim}\left(\left[\begin{array}{c|c} A & B \\ \hline C & D \end{array}\right]\right) \geq 0$ *exist such that $\sigma(Y\bar{Y}) < 1$.*

We investigate the properties of the Hamiltonian matrices (6.69), (6.70), (6.71), and (6.72) which play important roles in subsequent chapters. First of all, we note the relation between the zeros of $G(s)$ and the Hamiltonian matrices (6.69) and (6.71). Let λ be a zero of $G(s)$ given in (6.17); that is,

$$\begin{bmatrix} A-\lambda I & B \\ C & D \end{bmatrix}\begin{bmatrix} x \\ u \end{bmatrix} = 0, \tag{6.75}$$

for some x and u, when D is tall (more rows than columns), and

$$\begin{bmatrix} x^T & u^T \end{bmatrix} \begin{bmatrix} A - \lambda I & B \\ C & D \end{bmatrix} = 0 \tag{6.76}$$

for some x and u, when D is fat (more columns than rows). It is easy to see that (6.75) implies

$$f\left(\left[\begin{array}{c|c} A & B \\ \hline C & D \end{array}\right]\right) \begin{bmatrix} x \\ 0 \end{bmatrix} = \lambda \begin{bmatrix} x \\ 0 \end{bmatrix}$$

and (6.76) implies

$$f^{\sim}\left(\left[\begin{array}{c|c} A & B \\ \hline C & D \end{array}\right]\right) \begin{bmatrix} x \\ 0 \end{bmatrix} = \lambda \begin{bmatrix} x \\ 0 \end{bmatrix}.$$

Thus, we have established the following result.

LEMMA 6.15 *If $G(s)$ is tall, then each zero of $G(s)$ is an eigenvalue of $f\left(\left[\begin{array}{c|c} A & B \\ \hline C & D \end{array}\right]\right)$, and if $G(s)$ is fat, then each zero of $G(s)$ is an eigenvalue of $f^{\sim}\left(\left[\begin{array}{c|c} A & B \\ \hline C & D \end{array}\right]\right)$.*

Due to Theorem 3.2, we have the following important result.

LEMMA 6.16 *If $G(s)$ has a zero on the $j\omega$-axis, then it has no (J,J')-lossless factorization.*

The Hamiltonian matrices (6.69) and (6.71) have some invariance properties that can be proven directly.

LEMMA 6.17 *For any F, L, and invertible U and V, the following identities hold.*

$$f\left(\left[\begin{array}{c|c} A+BF & BU \\ \hline C+DF & DU \end{array}\right]\right) = f\left(\left[\begin{array}{c|c} A & B \\ \hline C & D \end{array}\right]\right), \tag{6.77}$$

$$f^{\sim}\left(\left[\begin{array}{c|c} A+LC & B+LD \\ \hline VC & VD \end{array}\right]\right) = f^{\sim}\left(\left[\begin{array}{c|c} A & B \\ \hline C & D \end{array}\right]\right). \tag{6.78}$$

Proof. The identity (6.77) easily follows from (6.69) and the relation

$$\begin{bmatrix} A+BF & 0 \\ -(C+DF)^T J(C+DF) & -(A+BF)^T \end{bmatrix}$$

$$= \begin{bmatrix} A & 0 \\ -C^T JC & -A^T \end{bmatrix}$$

$$- \begin{bmatrix} -B \\ (C+DF)^T JD \end{bmatrix} (D^T JD)^{-1} \begin{bmatrix} D^T JDF & 0 \end{bmatrix}$$

$$- \begin{bmatrix} 0 \\ F^T D^T JD \end{bmatrix} (D^T JD)^{-1} \begin{bmatrix} D^T JC & B^T \end{bmatrix}.$$

The identity (6.78) can be shown similarly.

If $G(s)$ is invertible, we have

$$(G^\sim(s)JG(s))^{-1} = G(s)^{-1}J(G^\sim(s))^{-1}.$$

Also, we have

$$(G(s)JG^\sim(s))^{-1} = (G^\sim(s))^{-1}JG(s)^{-1}.$$

From these identities and the definitions of f and g, we have the following result.

LEMMA 6.18 *If $G(s)$ is invertible, we have*

$$f\left(\left[\begin{array}{c|c} A & B \\ \hline C & D \end{array}\right]\right) = g^\sim\left(\left[\begin{array}{c|c} A & B \\ \hline C & D \end{array}\right]^{-1}\right), \tag{6.79}$$

$$f^\sim\left(\left[\begin{array}{c|c} A & B \\ \hline C & D \end{array}\right]\right) = g\left(\left[\begin{array}{c|c} A & B \\ \hline C & D \end{array}\right]^{-1}\right). \tag{6.80}$$

Notes

The J-lossless factorization, which is the subject of this chapter, is a strengthened notion of J-spectral factorization defined through (6.4).

It is sometimes called J-inner/outer factorization [3]. I am not sure people are aware of the importance of this factorization that includes inner/outer factorization of stable systems and spectral factorization of positive systems. The derivation of the J-lossless factorization based on J-lossless conjugation in this chapter is found in [58][59]. More rigorous treatment of J-lossless factorization is found in [26].

The procedure is divided into two stages, the extraction of zeros and that of poles. This is a very unique and clear algorithmic representation of the J-lossless factorization. Actually, the zeros extraction is equivalent to the cascade synthesis [38][104] of electrical circuits.

Problems

[1] Let $R > 0$ and

$$U(s) = \left[\begin{array}{c|c} A & B \\ \hline C & D \end{array}\right].$$

Find a condition on $U(s)$ such that

$$\Gamma(s) := R - U^{\sim}(s)JU(s)$$

is positive, that is, $\Gamma(j\omega) > 0$ for each ω.

[2] Obtain the condition under which both X and $\bar{X}$ are zero in Theorem 6.6. In that case, $\Theta(s)$ in (6.32) becomes the identity.

[3] Prove that if G is invertible, then G has a J-lossless factorization iff G^{-1} has a dual J-lossless factorization.

[4] Using the relation (6.21) which is written as $D^TJDF = -(D^TJC + B^TX)$, show that an augmentation (B_1, D_1) in (6.23) satisfies the condition (6.27) if

$$D_1^TJDF = -D_1^TJ(C + DF).$$

[5] Assume that $J - D(D^TJD)^{-1}D^T \geq 0$ in (6.19). Show that Ker$X$ is A-invariant; that is, AKer$X \subset$ Ker$X$.

[6] Compute the J_{11}-lossless factorization of the system

$$G(s) = \begin{bmatrix} \dfrac{2(s-1)}{s+2} & \dfrac{s-1}{s+1} \\ \dfrac{s+1}{s-2} & \dfrac{2(s+2)}{s-2} \end{bmatrix}.$$

[7] Assume that G is given in (6.17). Show that in Theorem 6.11,

$$GJ\Psi_+J = \left[\begin{array}{c|c} A - BJB^T\bar{Y} & B \\ \hline C - DJB^T\bar{Y} & D \end{array}\right],$$

where $\bar{Y}$ satisfies (iii) of Theorem 6.12. Also, show that $GJ\Psi_+J$ has a dual J-lossless factorization iff (ii) and (iv) of Theorem 6.12 hold.

[8] Let

$$M = Ric\ f\left(\left[\begin{array}{c|c} A & B \\ \hline C & D \end{array}\right]\right), \quad N = Ric\ g\left(\left[\begin{array}{c|c} A & B \\ \hline C & D \end{array}\right]\right).$$

Show that if $I - MN$ is invertible,

$$(I - MN)^{-1}M = Ric\ f\left(\left[\begin{array}{c|c} A + NC^TJC & B + NC^TJC \\ \hline C & D \end{array}\right]\right).$$

Chapter 7

H^{∞} Control via (J, J')-Lossless Factorization

7.1 Formulation of H^{∞} Control

Now we are in a position to deal with the H^{∞} control problem based on the results obtained in the previous chapters.

The plant to be controlled is described as

$$\begin{bmatrix} z \\ y \end{bmatrix} = P \begin{bmatrix} w \\ u \end{bmatrix} = \begin{bmatrix} P_{11} & P_{12} \\ P_{21} & P_{22} \end{bmatrix} \begin{bmatrix} w \\ u \end{bmatrix}, \tag{7.1}$$

where

$$\begin{aligned} z &: \text{ errors to be reduced } (\dim(z) = m), \\ y &: \text{ observation output } (\dim(y) = q), \\ w &: \text{ exogenous input } (\dim(w) = r), \\ u &: \text{ control input } (\dim(u) = p). \end{aligned}$$

A controller is given by

$$u = Ky. \tag{7.2}$$

The closed-loop system corresponding to this controller is shown in Figure 7.1. The closed-loop transfer function is given by

$$\Phi = P_{11} + P_{12}K(I - P_{22}K)^{-1}P_{21}. \tag{7.3}$$

Using the notation (4.80), Φ is written as

$$\Phi = LF\,(P; K). \tag{7.4}$$

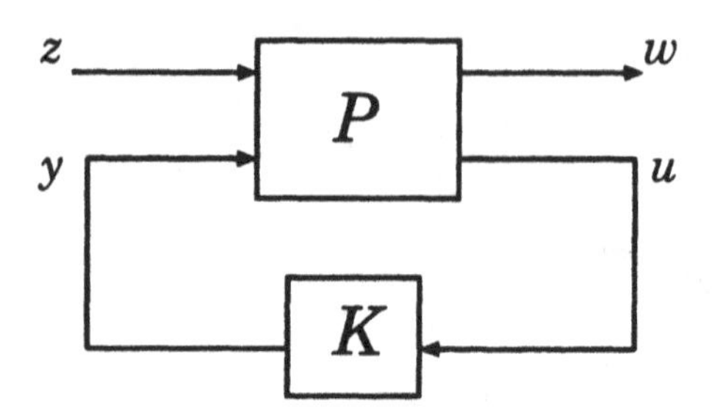

Figure 7.1 Closed-Loop System.

The H^∞ control problem is formulated as follows.

H^∞ Control Problem

Find a controller K such that the closed-loop system of Figure 7.1 *is internally stable and the closed-loop transfer function Φ given in* (7.4) *satisfies*

$$\|\Phi\|_\infty < \gamma \tag{7.5}$$

for a positive number $\gamma > 0$.

Several classical synthesis problems of practical importance can be reduced to the H^∞ control problem. These problems are concerned with the synthesis of a controller K of the unity feedback scheme described in Figure 7.2.

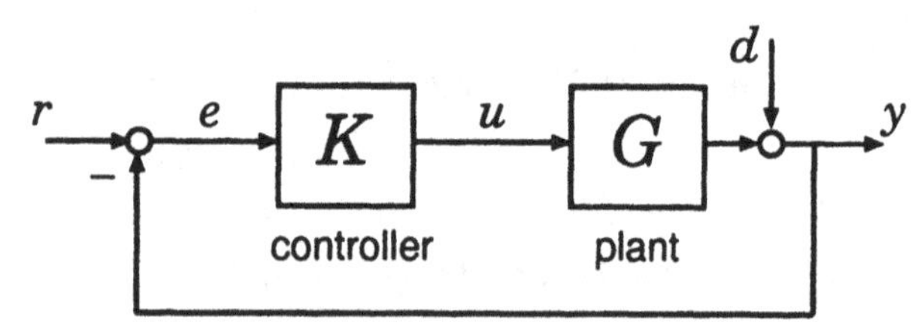

Figure 7.2 Unity Feedback Scheme.

[1] *Sensitivity Reduction Problem*

The objective is to reduce the magnitude of the *sensitivity function* S of Figure 7.2 given by

$$S = (I + GK)^{-1} \tag{7.6}$$

over a specified frequency range Ω. Note that S denotes the transfer function from r to e. Choosing an appropriate frequency weighting function $W(s)$ which is significant on $s = j\omega \in \Omega$, the problem is reduced to finding a controller K that stabilizes the closed-loop system of Figure 7.2 and satisfies

$$\|WS\|_\infty < \gamma \tag{7.7}$$

for some specified number $\gamma > 0$.

To reduce the synthesis problem specified by (7.7) to an H^∞ control problem, it is sufficient to choose a generalized plant P in Figure 7.1 for which Φ given in (7.3) coincides with WS. An example of such P is given by $P_{11} = W, P_{12} = -WG, P_{21} = I, P_{22} = -G$; that is,

$$P = \begin{bmatrix} W & -WG \\ I & -G \end{bmatrix}. \tag{7.8}$$

Thus, the sensitivity reduction problem specified by (7.7) is reduced to solving the H^∞ control problem for the plant given by (7.8).

[2] *Robust Stabilization Problems* [50][98]

Now, we consider the case where G in Figure 7.2 contains uncertainties in the sense that G is represented by

$$G(s) = G_0(s) + \Delta(s)W(s) \tag{7.9}$$

where $G_0(s)$ is a given nominal plant model, $W(s)$ is a given weighting function, and $\Delta(s)$ is an unknown function that is only known to be stable and satisfies

$$\|\Delta\|_\infty < 1. \tag{7.10}$$

The class of plants that can be represented in (7.9) is often referred to as the plant with *additive unstructured uncertainty.*

It is well known that a controller K stabilizes the closed-loop system of Figure 7.2 for all plants $G(s)$ described in (7.9) iff K stabilizes G_0 and satisfies

$$\|WQ\|_\infty < 1, \tag{7.11}$$

$$Q := K(I + G_0K)^{-1}. \tag{7.12}$$

This can be proven by observing the equivalence of Figure 7.3(a) and (b) and by appealing to the small gain theorem. The problem is again reduced to an H^∞ control problem by choosing P for which Φ given in (7.3) coincides with $WQ = WK(I + G_0K)^{-1}$. An example of such P is given by

$$P = \begin{bmatrix} 0 & W \\ I & -G_0 \end{bmatrix}. \tag{7.13}$$

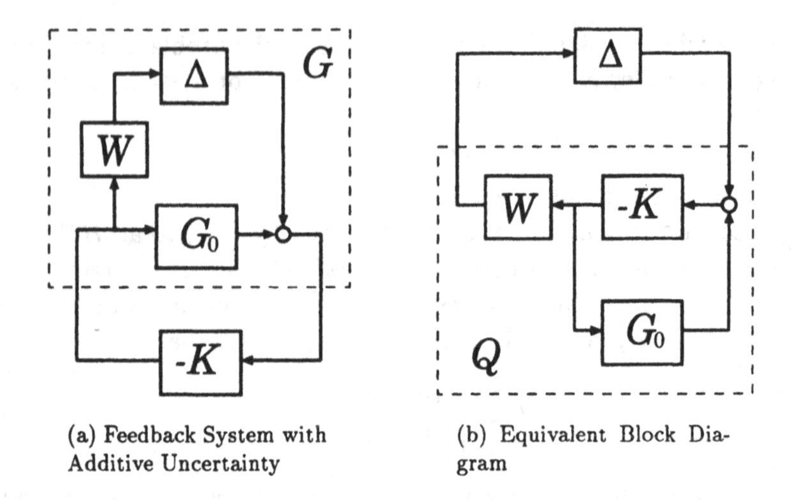

(a) Feedback System with Additive Uncertainty

(b) Equivalent Block Diagram

Figure 7.3 Robust Stabilization.

Instead of (7.9), we can represent the plant with uncertainty as

$$G = (I + \Delta(s)W(s))G_0(s), \tag{7.14}$$

where $\Delta(s)$ is a matrix representing uncertainty satisfying (7.10). This description is usually referred to as the plant with *multiplicative unstructured uncertainty.* In this case, the condition (7.11) is replaced by

$$\|WT\|_\infty < 1, \tag{7.15}$$

$$T := G_0K(I + G_0K)^{-1}. \tag{7.16}$$

The function T is often referred to as a *complementary sensitivity function.* The corresponding generalized plant is given by

$$P = \begin{bmatrix} 0 & WG_0 \\ I & -G_0 \end{bmatrix}. \tag{7.17}$$

[**3**] *Mixed Sensitivity Problem* [97]

The sensitivity S defined in (7.6) describes the transfer function from the disturbance d to the output y in Figure 7.2. Therefore, reduction of S implies disturbance rejection. On the other hand, as was stated previously, the reduction of the complementary sensitivity function T in (7.16) implies the enhancement of robust stability with respect to multiplicative uncertainty. Due to an obvious relation

$$S + T = I,$$

it is impossible to reduce both S and T simultaneously. Therefore, some sort of tradeoff in terms of frequency band between reduction of S and that of T is required. A method of achieving such a tradeoff is to consider a weighted sum of S and T as a performance criterion Φ; that is,

$$\Phi = \begin{bmatrix} W_1 S \\ W_2 T \end{bmatrix}, \tag{7.18}$$

where W_1 and W_2 are appropriately chosen weighting functions. The specification (7.5) is equivalently represented as

$$S^*(j\omega)R(j\omega)S(j\omega) + T^*(j\omega)Q(j\omega)T(j\omega) < \gamma^2 I, \tag{7.19}$$
$$(R(s) = W_1^T(-s)W_1(s), \quad Q(s) = W_2^T(-s)W_2(s))$$

for each ω. A generalized plant that generates Φ in (7.18) is given by

$$P = \left[\begin{array}{c|c} W_1 & -W_1 G_0 \\ 0 & W_2 G_0 \\ \hline I & -G_0 \end{array}\right]. \tag{7.20}$$

Instead of (7.18), we can take

$$\Phi = \begin{bmatrix} W_1 Q \\ W_2 T \end{bmatrix} \tag{7.21}$$

to consider the tradeoff between Q, and T rather than S and T. A generalized plant corresponding to (7.21) is given by

$$P = \left[\begin{array}{c|c} 0 & W_1 \\ 0 & W_2 G_0 \\ \hline I & -G_0 \end{array}\right]. \tag{7.22}$$

Furthermore, we can consider the tradeoffs among S, Q, and T by considering

$$\Phi = \begin{bmatrix} W_1 S \\ W_2 Q \\ W_3 T \end{bmatrix}, \tag{7.23}$$

which is generated by the generalized plant

$$P = \left[\begin{array}{c|c} W_1 & -W_1G_0 \\ 0 & W_2 \\ 0 & W_3G_0 \\ \hline I & -G_0 \end{array}\right]. \tag{7.24}$$

7.2 Chain-Scattering Representations of Plants and H^∞ Control

The H^∞ control problem can also be represented based on the chain-scattering representations of the plant.

We assume that P_{21} is square and invertible. This implies

$$q = r. \tag{7.25}$$

Corresponding to (4.3), the plant (7.1) is represented as

$$\begin{bmatrix} z \\ w \end{bmatrix} = G \begin{bmatrix} u \\ y \end{bmatrix},$$

where G is the chain-scattering representation of the plant P given by

$$\begin{aligned} G &= CHAIN\,(P) \\ &= \begin{bmatrix} P_{12} - P_{11}P_{21}^{-1}P_{22} & P_{11}P_{21}^{-1} \\ -P_{21}^{-1}P_{22} & P_{21}^{-1} \end{bmatrix} \in \mathbf{R}(s)^{(m+r)\times(p+r)}. \end{aligned} \tag{7.26}$$

Now the plant (7.1) is alternatively represented as

$$\begin{bmatrix} z \\ w \end{bmatrix} = G \begin{bmatrix} u \\ y \end{bmatrix} = \begin{bmatrix} G_{11} & G_{12} \\ G_{21} & G_{22} \end{bmatrix} \begin{bmatrix} u \\ y \end{bmatrix}. \tag{7.27}$$

Using the notation developed in Section 4.6, the closed-loop transfer function Φ given in (7.3) is represented as

$$\Phi = HM\,(G; K). \tag{7.28}$$

Figure 7.4 describes the closed-loop system in terms of the chain-scattering representation of P.

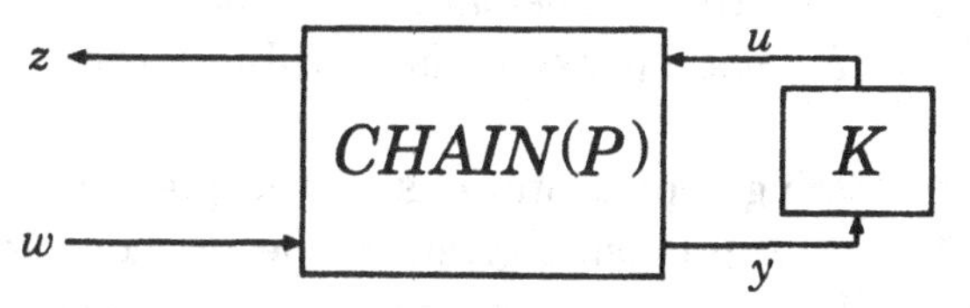

Figure 7.4 Chain-Scattering Representation of Closed-Loop System.

The H^∞ control problem is formulated in terms of the chain-scattering representation of the plant as follows.

H∞ Control Problem

Find a controller K such that the closed-loop system of Figure 7.4 is internally stable and

$$\|HM\,(G;K)\|_\infty < \gamma \tag{7.29}$$

for a given positive number $\gamma > 0$.

If P_{12}^{-1} exists instead of P_{21}^{-1}, which implies

$$m = p\,, \tag{7.30}$$

we can use the dual chain-scattering representation

$$\begin{aligned} H &= DCHAIN\,(P) \\ &= \begin{bmatrix} P_{12}^{-1} & -P_{12}^{-1}P_{11} \\ P_{22}P_{12}^{-1} & P_{21} - P_{22}P_{12}^{-1}P_{11} \end{bmatrix}. \end{aligned} \tag{7.31}$$

Then, due to (i) of Lemma 4.16, the closed-loop transfer function Φ in (7.4) is represented as

$$\Phi = DHM\,(H;K). \tag{7.32}$$

The H^∞ control problem is formulated in terms of the dual chain-scattering representation of the plant as follows:

H∞ Control Problem (Dual Formulation)

Find a controller K such that the closed-loop system of Figure 7.4 is internally stable and

$$\|DHM\,(H;K)\|_\infty < \gamma$$

for a given positive number $\gamma > 0$.

As was shown previously, when either P_{21} or P_{12} is invertible, we can use either the chain-scattering or the dual chain-scattering formalism. These cases correspond to the *one-block* and the *two-block* cases. The case where neither P_{21} nor P_{12} exist (the four-block case) is discussed in the next section.

The chain-scattering representations of the plant corresponding to the special H^∞ control problems given in the previous section can be computed easily. The plant (7.8) corresponding to the sensitivity reduction problem gives rise to the chain-scattering representation

$$CHAIN\,(P) = \begin{bmatrix} 0 & W \\ G & I \end{bmatrix},$$

which is much simpler than the original plant (7.8).

The plants (7.13) and (7.17) corresponding to the robust stabilization problems yield the chain-scattering representations

$$CHAIN\,(P) = \begin{bmatrix} W & 0 \\ G_o & I \end{bmatrix},$$

$$CHAIN\,(P) = \begin{bmatrix} WG_o & 0 \\ G_o & I \end{bmatrix},$$

respectively. The plants (7.20), (7.22), (7.24) which correspond to mixed sensitivity problems yield, respectively, the chain-scattering representations

$$CHAIN\,(P) = \left[\begin{array}{c|c} 0 & W_1 \\ W_2G_o & 0 \\ \hdashline G_o & I \end{array}\right],$$

$$CHAIN\,(P) = \left[\begin{array}{c|c} W_1 & 0 \\ W_2G_0 & 0 \\ \hdashline G_0 & I \end{array}\right],$$

$$CHAIN\,(P) = \left[\begin{array}{c|c} 0 & W_1 \\ W_2 & 0 \\ W_3G_0 & 0 \\ \hdashline G_0 & I \end{array}\right].$$

These representations show that the chain-scattering representation is sometimes simpler than the original plant description.

An important advantage of the chain-scattering representation is that it gives a possibility of simplifying the problem by factoring out a unimodular portion in the chain-scattering representation, as is shown in the following lemma.

LEMMA 7.1 *Assume that $G = CHAIN\,(P)$ is represented in the factored form*

$$G(s) = G_0(s)\Pi(s), \tag{7.33}$$

where $\Pi(s)$ is unimodular. Then the H^∞ control problem for the plant P is solvable iff the H^∞ control problem for the plant P_0 is solvable, where $G_0 = CHAIN\,(P_0)$.

Proof. Let K_0 be a solution to the H^∞ control problem for P_0 and let

$$K = HM\,(\Pi^{-1}; K_0). \tag{7.34}$$

Due to (iii) of Lemma 4.13, we have

$$HM\,(G; K) = HM\,(G_0\Pi; HM\,(\Pi^{-1}; K_0)) = HM\,(G_0; K_0).$$

Since K_0 solves the H^∞ control problem for P_0, $HM\,(G_0; K_0)$ is internally stable and $\|HM\,(G_0; K_0)\|_\infty < \gamma$. Hence, $\|HM\,(G; K)\|_\infty < \gamma$. Due to Lemma 4.15, $HM\,(G; K)$ is internally stable. Hence, (7.34) solves the H^∞ control problem for P. By the reverse argument, we can show that if K solves the H^∞ control problem for P, then $K_0 = HM\,(\Pi; K)$ solves the H^∞ control problem for P_0. ∎

Figure 7.5 illustrates the implication of Lemma 7.1. It should be noticed that the problem reduction from P to P_0 is ascribed to the cascade structure of the chain-scattering representation.

In the unity feedback system of Figure 7.6(a), the plant G is assumed to be decomposed into the product

$$G = G_0\Psi,$$

where Ψ is a unimodular system. Then the synthesis problem for G is reduced to that of a simpler system G_0 because, if K_0 is a desired controller for G_0, then $K = \Psi^{-1}K_0$ is a desired controller (at least from the input/output point of view) for G, as is shown in Figure 7.6(b). Lemma 7.1 generalizes this fact to more general schemes with disturbance.

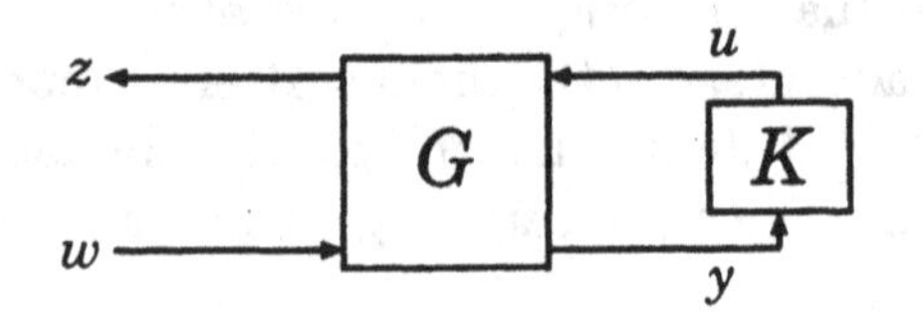

(a) Chain-Scattering Representation

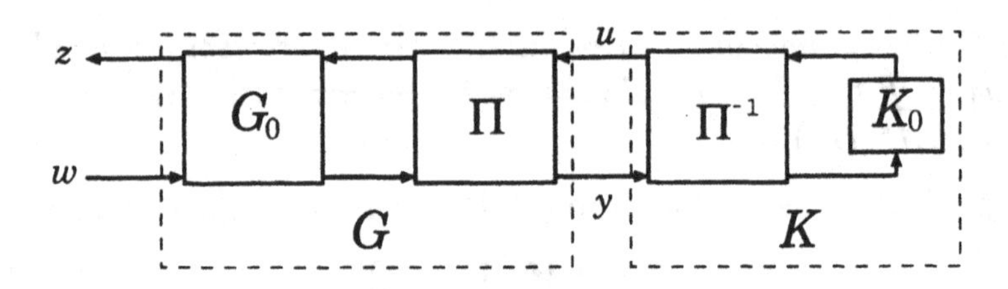

(b) Factorization of G and K

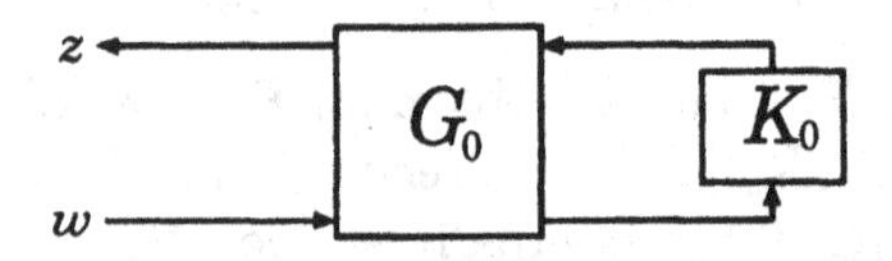

(c) Equivalent but Simpler Problem

Figure 7.5 Problem Reduction by Factorization of $CHAIN\,(P)$.

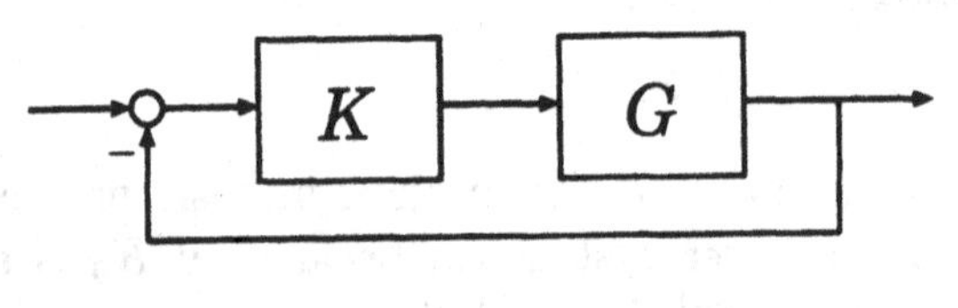

(a) A Unity-Feedback Scheme

Figure 7.6 Problem Reduction in Unity Feedback Systems.

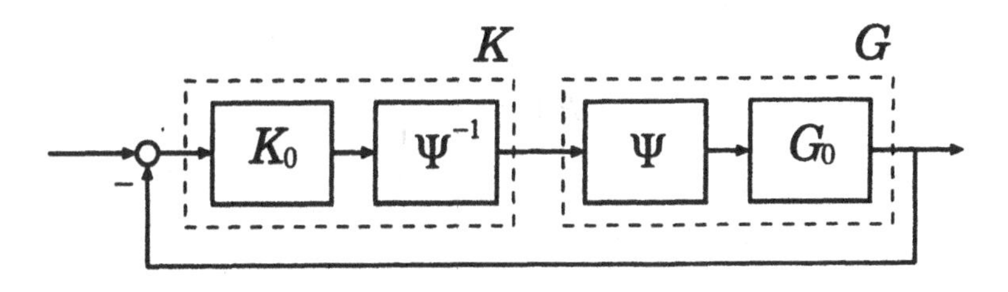

(b) Decomposition of G and K

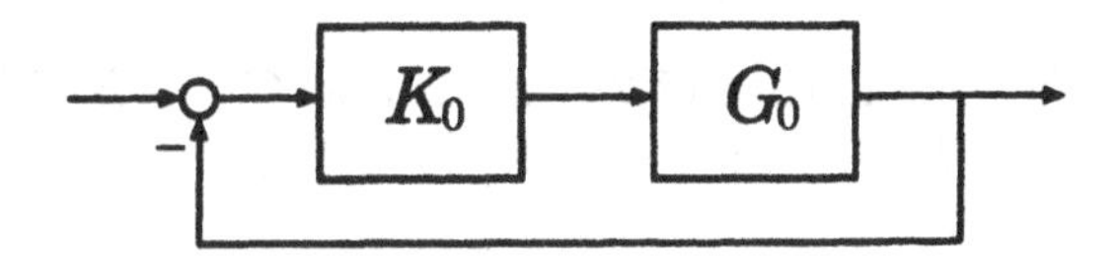

(c) Equivalent but Simpler Problem

Figure 7.6 continued.

7.3 Solvability Conditions for Two-Block Cases

In this section, we derive a necessary and sufficient condition for the solvability of the H^∞ control problem in terms of J-lossless factorization in the case where either a chain-scattering representation or its dual of the plant exists. This requires that the condition (7.25) which implies

$$\sharp(\text{observation outputs}) = \sharp(\text{exogenous inputs}),$$

or the condition (7.30) which implies

$$\sharp(\text{control inputs}) = \sharp(\text{controlled errors})$$

holds. In these cases, which are usually referred to as *two-block cases*, the solvability conditions are derived relatively easily. General cases are treated in Section 7.4.

Henceforth, we assume, without loss of generality, that

$$\gamma = 1 \tag{7.35}$$

in (7.5), by considering the normalized plants

$$P_\gamma = P \begin{bmatrix} \gamma^{-1} I & 0 \\ 0 & 1 \end{bmatrix} \tag{7.36a}$$

or

$$P'_\gamma = \begin{bmatrix} \gamma^{-1} I & 0 \\ 0 & 1 \end{bmatrix} P, \tag{7.36b}$$

instead of P. We assume that (7.25) holds and $G = CHAIN\,(P)$ exists. The problem is to find a controller K such that the closed system is internally stable and

$$HM\,(G; K) \in \mathbf{BH}_\infty^{m\times r}. \tag{7.37}$$

The special case where G is (J_{mr}, J_{pr})-lossless has already been solved in Theorem 4.15. In this case, Theorem 4.15 implies that any $K \in \mathbf{BH}_\infty^{p\times r}$ solves the problem. Therefore, according to Lemma 7.1, the problem is solvable for any G which is represented as

$$G = \Theta\Pi, \tag{7.38}$$

where Θ is (J_{mr}, J_{pr})-lossless and Π is unimodular. The representation (7.38) is exactly the (J, J')-lossless factorization that was extensively discussed in Chapter 6. Thus, we have established the following result.

LEMMA 7.2 *Assume that P has a chain-scattering representation $G = CHAIN\,(P)$ which has no poles nor zeros on the $j\omega$-axis. Then, the normalized H^∞ problem is solvable if G has a (J, J')-lossless factorization (7.38). A desired controller is given by*

$$K = HM\,(\Pi^{-1}; S), \tag{7.39}$$

where S is any matrix in $\mathbf{BH}_\infty^{p\times r}$.

Now we prove that the existence of a (J, J')-lossless factorization of G is also a necessary condition for the solvability of the H^∞ control problem under a mild condition. The following result is a crucial step towards this end.

LEMMA 7.3 *Assume that the chain-scattering representation $G = CHAIN\,(P)$ is left invertible and has no poles or zeros on the $j\omega$-axis. Then the H^∞ control problem is solvable for G only if there exists a unimodular matrix Π such that*

$$G^\sim JG = \Pi^\sim J'\Pi. \tag{7.40}$$

Proof. Assume that K solves the H^∞ control problem; that is,

$$\Phi = HM\,(G;K) \in \mathbf{BH}_\infty^{m\times r}.$$

Due to (v) of Lemma 4.17 (actually, its generalization given in Problem 4.4),

$$\Phi = DHM\,(G^\dagger;K),$$

where $G^\dagger$ is a left inverse of G; that is, $G^\dagger G = I$.

Let $K = UV^{-1} = \hat{V}^{-1}\hat{U}$ be stable coprime factorizations of K, and write

$$\Xi = \begin{bmatrix} U \\ V \end{bmatrix}, \quad \hat{\Xi} = \begin{bmatrix} \hat{V} & -\hat{U} \end{bmatrix}. \tag{7.41}$$

From the definition of HM and DHM, it follows that $\Phi = M_1M_2^{-1} = -\hat{M}_1^{-1}\hat{M}_2$, where

$$M := \begin{bmatrix} M_1 \\ M_2 \end{bmatrix} = G\Xi, \quad \hat{M} := \begin{bmatrix} \hat{M}_1 & \hat{M}_2 \end{bmatrix} = \hat{\Xi}G^\dagger.$$

Obviously,

$$\hat{M}M = \hat{\Xi}G^\dagger G\Xi = 0. \tag{7.42}$$

From $M^\sim JM = M_1^\sim M_1 - M_2^\sim M_2$, $\hat{M}J\hat{M}^\sim = \hat{M}_1\hat{M}_1^\sim - \hat{M}_2\hat{M}_2^\sim$, and $\|\Phi\|_\infty < 1$, we have

$$M^\sim JM < 0, \quad \hat{M}J\hat{M}^\sim > 0, \quad \forall \mathrm{Re}\, s \geq 0. \tag{7.43}$$

Choose W and E both stable such that

$$\mathrm{Ker}\begin{bmatrix} JG & \hat{M}^\sim \end{bmatrix} = \mathrm{Im}\begin{bmatrix} W \\ E \end{bmatrix},$$

that is

$$JGW + \hat{M}^\sim E = 0. \tag{7.44}$$

Let

$$Z := \begin{bmatrix} W & \hat{\Xi} \end{bmatrix}.$$

We show that Z is unimodular. Assume, contrary to the assertion, that Z has a zero s_1 with $\mathrm{Re}[s_1] \geq 0$. Then there exists a vector ξ_1 and ξ_2 such that

$$W\xi_1 + \hat{\Xi}\xi_2 = 0, \quad s = s_1.$$

Premultiplication of the this identity by JG yields

$$-\hat{M}^\sim E\xi_1 + JM\xi_2 = 0,$$

which implies that $\xi_1^\sim E^\sim \hat{M} J \hat{M}^\sim E\xi_1 = \xi_2^\sim M^\sim JM\xi_2$. Due to (7.43), we conclude that $\xi_1 = 0, \quad \xi_2 = 0$. Thus, we have proven that Z is unimodular.

Due to the spectral factorization theorem (Theorem 3.8), there exist unimodular matrices Ψ and $\hat{\Psi}$ such that

$$\begin{aligned} M^\sim JM &= \Xi^\sim G^\sim JG\Xi = -\Psi^\sim\Psi, \\ E^\sim \hat{M} J \hat{M}^\sim E &= W^\sim G^\sim JGW = \hat{\Psi}^\sim\hat{\Psi}. \end{aligned}$$

Here we used the assumption that G has no poles or zeros on the imaginary axis.

Due to (7.44) and (7.42),

$$W^\sim G^\sim JG\Xi = -E^\sim \hat{M} M = 0.$$

Therefore, it follows that

$$Z^\sim G^\sim JGZ = \begin{bmatrix} \hat{\Psi}^\sim\hat{\Psi} & 0 \\ 0 & -\Psi^\sim\Psi \end{bmatrix}.$$

Hence,

$$\Pi := \begin{bmatrix} \hat{\Psi} & 0 \\ 0 & \Psi \end{bmatrix} Z^{-1} \tag{7.45}$$

satisfies Equation (7.40). The proof is now complete.

Due to (7.40), $\Theta = G\Pi^{-1}$ satisfies

$$\Theta^\sim J\Theta = J'.$$

Hence, it is J-unitary. In other words, the H^∞ control problem is solvable only if $G = CHAIN\,(P)$ allows a factorization

$$G = \Theta\Pi,$$

where Θ is (J, J')-unitary and Π is unimodular. Due to Lemma 7.1, the H^∞ control problem is solvable only if the H^∞ control problem is solvable for P_o with $\Theta = CHAIN(P_o)$. Theorem 4.15 implies that Θ must be (J, J')-lossless. Thus, we have shown that the H^∞ control problem is solvable only if $G = CHAIN\,(P)$ allows a (J, J')-lossless factorization (7.37), under the condition of Lemma 7.3. This, together with Lemma 7.2, establishes the following important result.

THEOREM 7.4 *Assume that the plant P given in (7.1) has a chain-scattering representation $G = CHAIN(P)$ such that G is left invertible and has no poles or zeros on the $j\omega$-axis. Then the normalized H^∞ control problem is solvable for P, iff G has a (J_{mr}, J_{pr})-lossless factorization*

$$G = CHAIN(P) = \Theta\Pi. \tag{7.46}$$

In that case, K is a desired controller iff

$$K = HM\,(\Pi^{-1}; S) \tag{7.47}$$

for an $S \in \mathbf{BH}_\infty^{p\times r}$.

This theorem establishes a close tie between the H^∞ control problem and (J, J')-lossless factorization. Actually, it reduces the H^∞ control problem to a factorization problem (7.38). This situation is analogous to the fact that the LQG problem is reduced to the Wiener-Hopf factorization of a positive matrix.

The form of the H^∞ controller (7.47) reveals an important characteristic feature of the closed-loop system which is illustrated in Figure 7.7. It shows that *the unimodular portion Π of $CHAIN\,(P)$ is totally cancelled out by the controller.* The representation (7.38) implies that *all the unstable poles and zeros are absorbed in Θ.* Therefore, the H^∞ controller cancels out all the stable poles and zeros of $CHAIN\,(P)$ and takes care of only the unstable poles and zeros from the power point of view. This is a remarkable feature of H^∞ control systems.

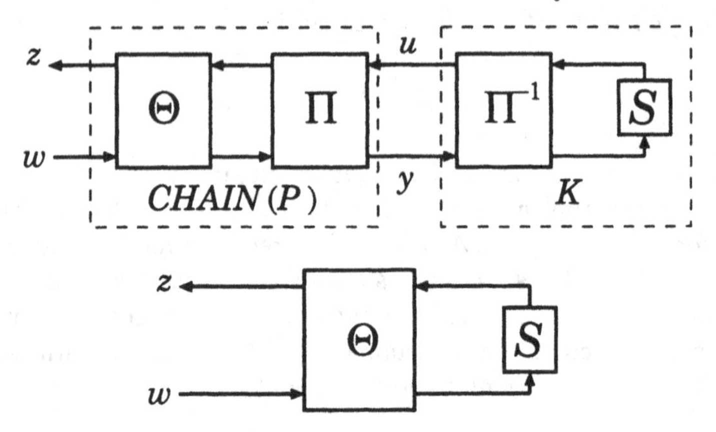

Figure 7.7 Closed-Loop Structure of H^∞ Control.

If *CHAIN* (P) has a stable pole or a stable zero close to the $j\omega$-axis, then it is cancelled out by a controller. This sort of cancellation might cause a poor closed-loop performance with respect to the initial conditions. This is a limitation of H^∞ control that only focuses on input/output behavior of the closed-loop system.

It is straightforward to dualize Theorem 7.4. We omit the detail and only state the result.

THEOREM 7.5 *Assume that the plant P given in (7.1) has a dual chain-scattering representation $H = DCHAIN\,(P)$ such that H is right-invertible and has no poles and zeros on the $j\omega$-axis. Then the normalized H^∞ control problem is solvable for P iff H has a dual (J_{mr}, J_{mq})-lossless factorization*

$$H = DCHAIN\,(P) = \Omega\Psi. \tag{7.48}$$

In that case, K is a desired controller iff

$$H = DHM\,(\Omega^{-1}; S) \tag{7.49}$$

for an $S \in \mathbf{BH}_\infty^{m\times q}$.

7.4 Plant Augmentations and Chain-Scattering Representations

In the preceding section, we assume that either P_{21} or P_{12} is invertible. If these assumptions fail, we cannot obtain the chain-scattering representations of the plant. In such cases, we augment the plant under the following conditions.

Assumption (A): There exist matrices P_{21}' and P_{12}' such that the inverses

$$\begin{bmatrix} P_{21} \\ P_{21}' \end{bmatrix}^{-1} := \begin{bmatrix} P_{21}^{\dagger} & P_{21}^{\perp} \end{bmatrix}, \quad \begin{bmatrix} P_{12}' & P_{12} \end{bmatrix}^{-1} = \begin{bmatrix} P_{12}^{\perp} \\ P_{12}^{\dagger} \end{bmatrix} \tag{7.50}$$

exist in $\mathbf{RL}^{\infty}$.
Obviously, (7.50) implies that

$$P_{21}P_{21}^{\dagger} = I, \quad P_{21}P_{21}^{\perp} = 0, \tag{7.51}$$

$$P_{12}^{\dagger}P_{12} = I, \quad P_{12}^{\perp}P_{12} = 0. \tag{7.52}$$

In other words, $P_{21}^{\dagger}$ and $P_{21}^{\perp}$ are a right inverse and a right annihilator of P_{21}, respectively, and $P_{12}^{\dagger}$ and $P_{12}^{\perp}$ are a left inverse and a left annihilator of P_{12}, respectively.

Now we consider the plant (7.1) with a fictitious observation output of dimension $r - q$ given by

$$y' = P_{21}'w + P_{22}'u, \tag{7.53}$$

where P_{21}' is chosen to satisfy (7.50). The augmented plant is described by

$$\begin{bmatrix} z \\ y \\ y' \end{bmatrix} = P_o \begin{bmatrix} w \\ u \end{bmatrix} = \left[\begin{array}{c|c} P_{11} & P_{12} \\ \hline P_{21} & P_{22} \\ P_{21}' & P_{22}' \end{array}\right] \begin{bmatrix} w \\ u \end{bmatrix}. \tag{7.54}$$

We obtain the chain-scattering representation of P_o as

$$\begin{bmatrix} z \\ w \end{bmatrix} = G \begin{bmatrix} u \\ y \\ y' \end{bmatrix} \tag{7.55}$$

$$
\begin{aligned}
G &= CHAIN(P_o) \\
&= \begin{bmatrix} P_{12} & P_{11} \\ 0 & I \end{bmatrix} \begin{bmatrix} I & 0 \\ P_{22} & P_{21} \\ P'_{22} & P'_{21} \end{bmatrix}^{-1} \\
&= \begin{bmatrix} P_{12} - P_{11}(P_{21}^{\dagger}P_{22} + P_{21}^{\perp}P'_{22}) & P_{11}P_{21}^{\dagger} & P_{11}P_{21}^{\perp} \\ -(P_{21}^{\dagger}P_{22} + P_{21}^{\perp}P'_{22}) & P_{21}^{\dagger} & P_{21}^{\perp} \end{bmatrix}.
\end{aligned} \tag{7.56}
$$

Since the controller (7.2) is represented as

$$u = \begin{bmatrix} K & 0 \end{bmatrix} \begin{bmatrix} y \\ y' \end{bmatrix}, \tag{7.57}$$

the closed-loop transfer function Φ from w to z is calculated to be

$$\Phi = \begin{bmatrix} G_{11}K + G_{12} & G_{13} \end{bmatrix} \begin{bmatrix} G_{21}K + G_{22} & G_{23} \end{bmatrix}^{-1}, \tag{7.58}$$

where G in (7.55) is represented in the block partitioned form

$$G = \begin{bmatrix} G_{11} & G_{12} & G_{13} \\ G_{21} & G_{22} & G_{23} \end{bmatrix} \begin{matrix} m \\ r \end{matrix} . \tag{7.59}$$
$$\qquad \quad p \qquad q \qquad r-q$$

According to the definition of HM, we can write Φ in (7.58) alternatively as

$$\Phi = HM\left(G\,;\begin{bmatrix} K & 0 \end{bmatrix}\right).$$

The closed-loop system corresponding to the controller (7.58) is shown in Figure 7.8.

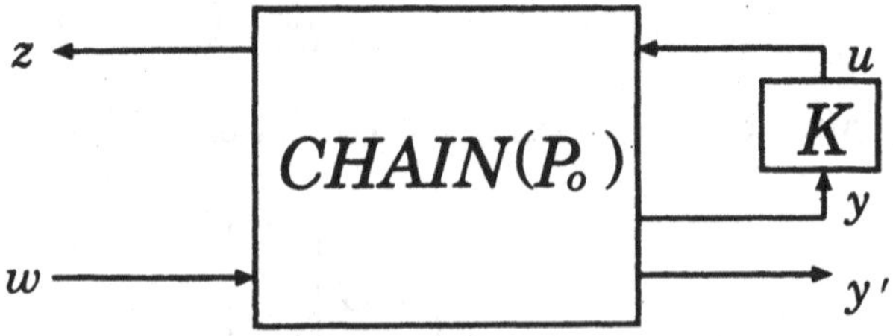

Figure 7.8 Closed-loop System with Output Augmentation.

In order to obtain the dual chain-scattering representation of P in (7.1), we consider a fictitious control input u' and augment P as follows :

$$z = P_{11}w + P'_{12}u' + P_{12}u \tag{7.60a}$$

$$y = P_{21}w + P'_{22}u' + P_{22}u, \tag{7.60b}$$

where P'_{12} is chosen to satisfy (7.50). The augmented plant is denoted by P_i; that is,

$$\begin{bmatrix} z \\ y \end{bmatrix} = P_i \begin{bmatrix} w \\ u' \\ u \end{bmatrix} = \left[\begin{array}{c|cc} P_{11} & P'_{12} & P_{12} \\ \hline P_{21} & P'_{22} & P_{22} \end{array}\right] \begin{bmatrix} w \\ u' \\ u \end{bmatrix}. \tag{7.60c}$$

Now we obtain, from (4.28), the dual chain-scattering representation of P_i as

$$\begin{bmatrix} u' \\ u \\ y \end{bmatrix} = H \begin{bmatrix} z \\ w \end{bmatrix}, \tag{7.61}$$

$$\begin{aligned} H &= DCHAIN\,(P_i) \\ &= \begin{bmatrix} P'_{12} & P_{12} & 0 \\ P'_{22} & -P_{22} & I \end{bmatrix}^{-1} \begin{bmatrix} I & -P_{11} \\ 0 & P_{21} \end{bmatrix} \\ &= \begin{bmatrix} P_{12}^{\perp} & -P_{12}^{\perp}P_{11} \\ P_{12}^{\dagger} & -P_{12}^{\dagger}P_{11} \\ P'_{22}P_{12}^{\perp} + P_{22}P_{12}^{\dagger} & P_{21} - (P'_{22}P_{12}^{\perp} + P_{22}P_{12}^{\dagger})P_{11} \end{bmatrix}, \end{aligned} \tag{7.62}$$

where $P_{12}^{\perp}$ and $P_{12}^{\dagger}$ are a left annihilator and a left inverse of P_{12} satisfying (7.52). Write H in a partitioned form conformable with (7.61) as

$$H = \begin{bmatrix} H_{11} & H_{12} \\ H_{21} & H_{22} \\ H_{31} & H_{32} \end{bmatrix}. \tag{7.63}$$

Due to (7.61) and (7.2), the closed-loop transfer function (7.3) with controller (7.2) is denoted as

$$\Phi = DCHAIN\left(H ; \begin{bmatrix} 0 \\ K \end{bmatrix}\right)$$
$$= -\begin{bmatrix} H_{11} \\ H_{21} - KH_{31} \end{bmatrix}^{-1} \begin{bmatrix} H_{12} \\ H_{22} - KH_{32} \end{bmatrix}.$$

Figure7.9 illustrates the closed-loop system in chain-scattering formalism.

Figure 7.9 Closed-Loop System with Input Augmentation.

Let us denote the third column block of G by

$$H^\perp := \begin{bmatrix} G_{13} \\ G_{23} \end{bmatrix} = \begin{bmatrix} P_{11} \\ I \end{bmatrix} P_{21}^\perp. \tag{7.64}$$

This notation comes from the fact that

$$HH^\perp = 0, \tag{7.65}$$

which is obvious from the definition (7.62) of H .

Also, we denote the top row block of H by

$$G^\perp := \begin{bmatrix} H_{11} & H_{12} \end{bmatrix} = P_{12}^\perp \begin{bmatrix} I & -P_{11} \end{bmatrix}. \tag{7.66}$$

Analogous to (7.65), we have

$$G^\perp G = 0. \tag{7.67}$$

Actually, we can prove that

$$H \cdot G = \begin{bmatrix} 0 & 0_{(r-q)\times(m-p)} \\ I_{p+q} & 0 \end{bmatrix}. \tag{7.68}$$

Due to (7.3) and (7.51), we have the identity

$$\Phi P_{21}^{\perp} = P_{11} P_{21}^{\perp}$$

for any K. Therefore, if $\|\Phi\|_\infty < 1$, we have

$$(P_{21}^{\perp})^{\sim}(I - P_{11}^{\sim} P_{11}) P_{21}^{\perp} = (P_{21}^{\perp})^{\sim}(I - \Phi^{\sim}\Phi) P_{21}^{\perp} > 0, \quad \forall s = j\omega. \tag{7.69}$$

Analogously, we have the dual version of (7.69) as

$$P_{12}^{\perp}(I - P_{11} P_{11}^{\sim})(P_{12}^{\perp})^{\sim} > 0, \quad \forall s = j\omega. \tag{7.70}$$

These inequalities can be rewritten in terms of $H^{\perp}$ and $G^{\perp}$ yielding the following interesting result.

LEMMA 7.6 *If the H^∞ control problem is solvable, then, for any augmentations, we have*

$$(H^{\perp})^{\sim} J H^{\perp} < 0, \qquad G^{\perp} J (G^{\perp})^{\sim} > 0, \quad \forall s = j\omega. \tag{7.71}$$

Proof. The inequalities (7.71) are obtained by rewriting (7.69) and (7.70) using (7.64) and (7.66), respectively. ∎

The following result is a generalization of Theorem 7.4 to four-block cases.

THEOREM 7.7 *Assume that there exists an output augmentation (7.53) such that $G = CHAIN\,(P_o)$ has no zeros or poles on the $j\omega$-axis. Then the H^∞ control problem is solvable for the plant (7.1) iff there exists an output augmentation (7.53) such that its chain scattering representation $G = CHAIN\,(P_o)$ has a (J_{mr}, J_{pr})-lossless factorization (7.38) with Π being of the lower triangular form:*

$$\Pi = \begin{bmatrix} \Pi_{11} & 0 \\ \Pi_{21} & \Pi_{22} \end{bmatrix} \begin{matrix} \}\, p+q \\ \}\, r-q \end{matrix} \quad . \tag{7.72}$$
$$\qquad \underbrace{\phantom{\Pi_{11}}}_{p+q} \; \underbrace{\phantom{\Pi_{22}}}_{r-q}$$

If these conditions hold, a desirable H^∞ controller is given by

$$K = HM\,(\Pi_{11}^{-1}\,; S), \tag{7.73}$$

where S is an arbitrary matrix in $\mathbf{BH}^\infty_{(p+q)\times(p+q)}$. The closed-loop transfer function is given by

$$\Phi = HM\left(\Theta; \begin{bmatrix} S & 0 \end{bmatrix}\right). \tag{7.74}$$

Proof. To prove the sufficiency, it is sufficient to see that the controller (7.73) satisfies

$$\Phi = HM\ \left(G\,; \begin{bmatrix} K & 0 \end{bmatrix}\right) \in \mathbf{BH}^\infty,$$

under the condition that G has a (J, J')-lossless factorization (7.38) with Π being of the form (7.72). This is obvious from

$$\begin{aligned} HM\,(G; \begin{bmatrix} K & 0 \end{bmatrix}) &= HM\left(\Theta \begin{bmatrix} \Pi_{11} & 0 \\ \Pi_{21} & \Pi_{22} \end{bmatrix}; \begin{bmatrix} HM\,(\Pi_{11}^{-1}\,; S) & 0 \end{bmatrix}\right) \\ &= HM\left(\Theta \begin{bmatrix} \Pi_{11} & 0 \\ \Pi_{21} & \Pi_{22} \end{bmatrix};\right. \\ &\qquad\qquad \left. HM\left(\begin{bmatrix} \Pi_{11}^{-1} & 0 \\ 0 & I \end{bmatrix}; [\,S \ \ 0\,]\right)\right) \\ &= HM\left(\Theta \begin{bmatrix} I & 0 \\ \Pi_{21}\Pi_{11}^{-1} & \Pi_{22} \end{bmatrix}; [\,S \ \ 0\,]\right) \\ &= HM\,(\Theta; [\,S \ \ 0\,]) \in \mathbf{BH}^\infty, \end{aligned}$$

where we used the relations given in Problem 4.5.

To prove the necessity, we modify slightly the proof of Lemma 7.3. Since K is replaced by $\begin{bmatrix} K & 0 \end{bmatrix}$, we must replace Ξ and $\hat{\Xi}$ in (7.41) by

$$\Xi = \begin{bmatrix} U & 0 \\ V & 0 \\ 0 & I \end{bmatrix}, \quad \hat{\Xi} = \begin{bmatrix} \hat{V} & -\hat{U} & 0 \\ 0 & 0 & I \end{bmatrix}.$$

Then we conclude that Π given in (7.45) is of the lower triangular form (7.72). ∎

Let us consider the meaning of the form (7.72). If the conditions of Theorem 7.7 hold, we can represent the plant in Figure 7.10 where

$$\begin{bmatrix} z \\ w \end{bmatrix} = \begin{bmatrix} \Theta_{11} & \Theta_{12} & \Theta_{13} \\ \Theta_{21} & \Theta_{22} & \Theta_{23} \end{bmatrix} \begin{bmatrix} u_a \\ y_a \\ y_a' \end{bmatrix} \tag{7.75}$$

$$\begin{bmatrix} u_a \\ y_a \\ y_a' \end{bmatrix} = \left[\begin{array}{cc|c} \Pi_{11a} & \Pi_{11b} & 0 \\ \Pi_{11c} & \Pi_{11d} & 0 \\ \hline \Pi_{21a} & \Pi_{21b} & \Pi_{22} \end{array}\right] \begin{bmatrix} u \\ y \\ y' \end{bmatrix}. \tag{7.76}$$

The controller (7.73) implies that

$$\begin{bmatrix} u \\ y \end{bmatrix} = \begin{bmatrix} \Pi_{11a} & \Pi_{11b} \\ \Pi_{11c} & \Pi_{11d} \end{bmatrix}^{-1} \begin{bmatrix} S \\ I \end{bmatrix} v$$

for some signal v. Substituting the preceding relation in (7.75) yields

$$\begin{bmatrix} u_a \\ y_a \\ y_a' \end{bmatrix} = \begin{bmatrix} S & 0 \\ I & 0 \\ 0 & I \end{bmatrix} \begin{bmatrix} v \\ y_a' \end{bmatrix}.$$

Hence, we have, from (7.75),

$$\begin{bmatrix} z \\ w \end{bmatrix} = \begin{bmatrix} \Theta_{11} & \Theta_{12} & \Theta_{13} \\ \Theta_{21} & \Theta_{22} & \Theta_{23} \end{bmatrix} \begin{bmatrix} S & 0 \\ I & 0 \\ 0 & I \end{bmatrix} \begin{bmatrix} v \\ y_a' \end{bmatrix}.$$

Eliminating v and y_a' from this relation yields (7.74).

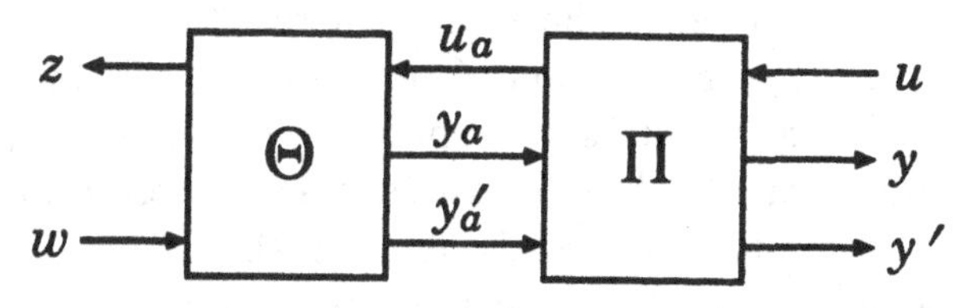

Figure 7.10 Plant Representation.

The dual of Theorem 7.7 is as follows.

THEOREM 7.8 *Assume that there exists an input augmentation (7.60) such that $H = DCHAIN(P_i)$ has no zeros or poles on the $j\omega$-axis. Then the H^∞ control problem is solvable for the plant (7.1) iff there exists an input augmentation (7.60) such that its dual chain-scattering representation $H = DCHAIN(P_i)$ has a dual (J_{mr}, J_{mq}) factorization (7.48) with Ω being of the lower triangular form*

$$\Omega = \underbrace{\begin{bmatrix} \Omega_{11} & 0 \\ \Omega_{21} & \Omega_{22} \end{bmatrix}}_{m-p \quad p+q} \begin{matrix} \} m-p \\ \} p+q \end{matrix} . \tag{7.77}$$

If these conditions hold, a desirable H^∞ controller is given by

$$K = DHM(\Omega_{22}^{-1}; S),$$

where S is an arbitrary matrix in $\mathbf{BH}^\infty_{(p+q)\times(p+q)}$. The closed-loop transfer function is given by

$$\Phi = DHM\left(\Psi; \begin{bmatrix} 0 \\ S \end{bmatrix}\right).$$

Notes

This chapter gave the formulation of the H^∞ control problem and established a close tie between H^∞ control and J-lossless factorization. The main results were Theorem 7.4 and Theorem 7.7 which gave the complete solvability conditions of the H^∞ control problem for the case where the plant has a chain-scattering representation and for the case where it doesn't, respectively. The equivalent results were obtained by Ball and Cohen [3], Green [35], and Green et al. [36], but their results were mainly stated in terms of J-spectrum factorization (7.40). In order to guarantee that $G\Pi^{-1}$ is J-lossless, they impose an additional condition on Π. A similar result without using the notion of J-lossless factorization was given in [60], which is based simply on the notion of the J-lossless complement as discussed in Chapter 4. Through the work of this chapter, it became clear that the H^∞ control problem is reduced to the J-lossless factorization. It is interesting to compare this result with the parallel fact in LQG control that the LQG control problem is reduced to the Wiener-Hopf factorization of positive matrices. As was discussed in Section 6.1, the J-lossless factorization includes the Wiener-Hopf factorization of positive matrices as a special case. In this sense, LQG control is a special case of H^∞ control.

Theorems 7.4 and 7.7 imply that the H^∞ controller must cancel out the unimodular part of the chain-scattering representation of the plant. In other words, all the stable zeros and poles of the chain-scattering representation are cancelled out. This is a significant characteristic feature of H^∞ control. This fact was pointed out by several researchers [63][72][71][88].

Problems

[1] Show that in the mixed sensitivity problem, a zero of G_0 is also a zero of the chain-scattering representation of $CHAIN\,(P)$, where P is given by (7.20) or (7.22).

[2] Prove that a zero of P is also a zero of P_o given by (7.54). Obtain the dual result.

[3] Assume that P_{21}^{-1} exists in (7.1). Show that the zeros of $CHAIN\,(P)$ are composed of the poles of P_{21} and the zeros of P_{12}.

[4] Assume that in the unity feedback system of Figure 7.2, the plant G is decomposed into

$$G = G_1 G_2,$$

where G_2 is unimodular. Show that the problem of robust stabilization and the mixed sensitivity problem for G are reduced to those of G_1 using the chain-scattering representation of G.

[5] Show that $CHAIN\,(P)$ is left invertible iff P_{12} is left invertible. Prove that the left inverse of $CHAIN\,(P)$ is given by

$$(CHAIN\,(P))^\dagger = \begin{bmatrix} P_{12}^\dagger & -P_{12}^\dagger P_{11} \\ P_{22}P_{12}^\dagger & P_{21} - P_{22}P_{12}^\dagger P_{11} \end{bmatrix}.$$

Compare this form with (7.31). Obtain the dual result.

[**6**] Show that we can perfectly reject the disturbance (i.e., we can find a stabilizing controller K that satisfies (7.5) for arbitrarily small γ) if $CHAIN\,(P)$ is unimodular.

[**7**] Compute, if possible, the dual chain-scattering representations of the transposes of the plants (7.8), (7.13), (7.17), (7.20), (7.22), and (7.24).

Chapter 8

State-Space Solutions to H^∞ Control Problems

8.1 Problem Formulation and Plant Augmentation

In this chapter, we give a solution to the H^∞ control problem in the state space based on the results obtained in the preceding chapters. A state-space realization of the plant (7.1) is given by

$$\dot{x} = Ax + B_1 w + B_2 u, \tag{8.1a}$$
$$z = C_1 x + D_{11} w + D_{12} u, \tag{8.1b}$$
$$y = C_2 x + D_{21} w + D_{22} u \tag{8.1c}$$

where

$z \in R^m$: errors to be reduced,
$y \in R^q$: observation outputs,
$w \in R^r$: exogenous inputs,
$u \in R^p$: control inputs.

We can write

$$P = \begin{bmatrix} P_{11} & P_{12} \\ P_{21} & P_{22} \end{bmatrix} = \left[\begin{array}{c|cc} A & B_1 & B_2 \\ \hline C_1 & D_{11} & D_{12} \\ C_2 & D_{21} & 0 \end{array}\right]. \tag{8.2}$$

Compared with the usual state-space description (4.17), we assume $D_{22} = 0$. This assumption holds for almost all plants of practical importance. We make the usual assumption that

(A1) (A, B_2) is stabilizable and (A, C_2) is detectable.

This assumption is necessary in order that the H^∞ control problem is solvable.

In this chapter, we deal with the so-called *standard problem* in which the following assumptions hold.

(A2) rank $P_{21}(j\omega) = q$, rank $P_{12}(j\omega) = p$, $0 \leq \omega \leq \infty$.

This assumption is equivalent to

(A2$_1$)

$$\text{rank} \begin{bmatrix} A - j\omega I & B_1 \\ C_2 & D_{21} \end{bmatrix} = n + q, \quad \forall\omega$$

(A2$_2$)

$$\text{rank} \begin{bmatrix} A - j\omega I & B_2 \\ C_1 & D_{12} \end{bmatrix} = n + p, \quad \forall\omega$$

(A2$_3$) rank $D_{21} = q$, rank $D_{12} = p$.

It should be noted that (A2$_3$) implies that

$$r \geq q, \quad m \geq p. \tag{8.3}$$

As was stated in Chapter 7, H^∞ control problems are usually categorized into the following classes.

(i) *One-block cases* where the equalities $r = q$, $m = p$ hold.

(ii) *Two-block cases* where either $r = q$ or $m = p$ holds.

(iii) *Four-block cases* where $r > q$ and $m > p$.

Since $D_{21} \in \mathbf{R}^{q \times r}$, $D_{12} \in \mathbf{R}^{m \times p}$, the assumption (A2$_3$) implies that both D_{21} and D_{12} are invertible in the one-block case, and either D_{21} or D_{12} is invertible in the two-block case. In the four-block case, neither D_{21} nor D_{12} are invertible.

From the viewpoint of the chain-scattering approach, H^∞ control problems are categorized as:

(i) one-block cases where both *CHAIN* (P) and *DCHAIN* (P) exist,

(ii) two-block cases where either *CHAIN* (P) or *DCHAIN* (P) exists,

(iii) four-block cases where neither *CHAIN* (P) nor *DCHAIN* (P) exists.

The cases (i) and (ii) are easily treated based on the results obtained in Chapters 6 and 7. In this case, the problem is reduced to the (J, J')–lossless factorization of *CHAIN* (P) and/or *DCHAIN* (P), as was discussed in Section 7.3. In these cases, we can obtain the solution to the problem by applying the results of the previous chapters in a straightforward way.

The case that needs further consideration is (iii) where we need to augment the plant to derive the chain-scattering representations of P. Corresponding to (7.54), we introduce a fictitious output

$$y' = C_2{}'x + D_{21}{}'w \tag{8.4}$$

in addition to (8.1c). We choose $D_{21}' \in \mathbf{R}^{(r-q)\times r}$ such that D_{21}' is a complement of D_{21} in the sense that

$$\hat{D}_{21} := \begin{bmatrix} D_{21} \\ D_{21}' \end{bmatrix}$$

is invertible. Hereafter, an augmented matrix is designated by "$\wedge$". A state-space realization of the output augmented plant P_o in (7.54) is given by

$$P_o = \left[\begin{array}{c|cc} A & B_1 & B_2 \\ \hline C_1 & D_{11} & D_{12} \\ C_2 & D_{21} & 0 \\ C_2' & D_{21}' & 0 \end{array}\right] = \left[\begin{array}{c|cc} A & B_1 & B_2 \\ \hline C_1 & D_{11} & D_{12} \\ \hat{C}_2 & \hat{D}_{21} & 0 \end{array}\right]. \tag{8.5}$$

Since $\hat{D}_{21}$ is invertible, *CHAIN* (P_o) exists and is given by

$$\begin{aligned} G &:= CHAIN\,(P_o) \\ &= \left[\begin{array}{c|cc} A - B_1\hat{D}_{21}^{-1}\hat{C}_2 & B_2 & B_1\hat{D}_{21}^{-1} \\ \hline C_1 - D_{11}\hat{D}_{21}^{-1}\hat{C}_2 & D_{12} & D_{11}\hat{D}_{21}^{-1} \\ -\hat{D}_{21}^{-1}\hat{C}_2 & 0 & \hat{D}_{21}^{-1} \end{array}\right] := \left[\begin{array}{c|c} A_G & B_G \\ \hline C_G & D_G \end{array}\right]. \end{aligned} \tag{8.6}$$

The preceding representation is obtained from (4.19) applied to (8.5).

Now, we apply the relation (4.7) to P_o in order to obtain a factorized form of G. For this purpose, we introduce "partial plants" given as

$$P_z := \begin{bmatrix} P_{11} & P_{12} \\ I & 0 \end{bmatrix} = \left[\begin{array}{c|c} A & B \\ \hline C_z & D_z \end{array}\right], \tag{8.7}$$

$$\hat{P}_y := \begin{bmatrix} 0 & I \\ \hat{P}_{21} & \hat{P}_{22} \end{bmatrix} = \left[\begin{array}{c|c} A & B \\ \hline \hat{C}_y & \hat{D}_y \end{array}\right], \tag{8.8}$$

where

$$B := \begin{bmatrix} B_1 & B_2 \end{bmatrix}, \; C_z := \begin{bmatrix} C_1 \\ 0 \end{bmatrix}, \; D_z := \begin{bmatrix} D_{11} & D_{12} \\ I & 0 \end{bmatrix}, \tag{8.9}$$

$$\hat{C}_y := \begin{bmatrix} 0 \\ \hat{C}_2 \end{bmatrix} = \begin{bmatrix} 0 \\ C_2 \\ C_2' \end{bmatrix}, \; \hat{D}_y := \begin{bmatrix} 0 & I \\ \hat{D}_{21} & 0 \end{bmatrix} = \begin{bmatrix} 0 & I \\ D_{21} & 0 \\ D_{21}' & 0 \end{bmatrix}. \tag{8.10}$$

These *partial plants* P_z and $\hat{P}_y$ correspond to $\Sigma_{1\cdot}$ and $\Sigma_{2\cdot}$ given in (4.6), respectively. The subscripts z in (8.7) and y in (8.8) represent the signals which are represented completely by the corresponding subplants. The identity (4.7) yields

$$\begin{aligned} G &= CHAIN\,(P_o) \\ &= CHAIN\,(P_z) \cdot CHAIN\,(\hat{P}_y). \end{aligned} \tag{8.11}$$

It is worth noting that the first factor $CHAIN\,(P_z)$ does not depend on the augmentation (C_2', D_{21}'). According to Lemma 4.1, we have an alternative representation of G in (8.6) as

$$G = \left[\begin{array}{c|c} A_G & B_G \\ \hline C_G & D_G \end{array}\right] = \left[\begin{array}{c|c} A - B\hat{D}_y^{-1}\hat{C}_y & B\hat{D}_y^{-1} \\ \hline C_z - D_z\hat{D}_y^{-1}\hat{C}_y & D_z\hat{D}_y^{-1} \end{array}\right]. \tag{8.12}$$

This representation implies that $G = CHAIN\,(P_o)$ is obtained by applying a state feedback gain $-\hat{D}_y^{-1}\hat{C}_y$ and an input transformation with $\hat{D}_y^{-1}$ as its transformation matrix to a partial plant P_z which does not depend on the augmentation (C_2', D_{21}').

Figure 8.1 illustrates the block diagram of $G = CHAIN\,(P_o)$.

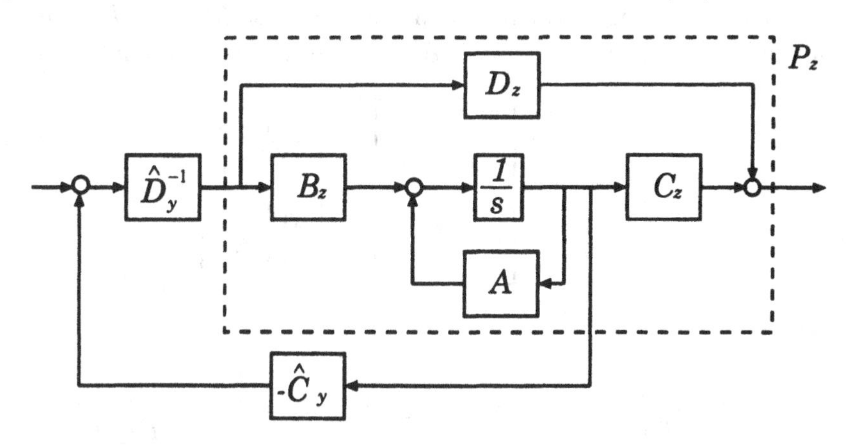

Figure 8.1 Block Diagram of $CHAIN\,(P_o)$.

Dually, we consider an input augmentation (7.60). A fictitious input u' is introduced as

$$\begin{aligned} \dot{x} &= Ax + B_1 w + B_2' u' + B_2 u, \\ z &= C_1 x + D_{11} w + D_{12}' u' + D_{12} u. \end{aligned}$$

The matrix $D_{12}' \in \mathbf{R}^{m \times (m-p)}$ is chosen such that it is a complement of D_{12} in the sense that

$$\hat{D}_{12} = \begin{bmatrix} D_{12}' & D_{12} \end{bmatrix}$$

is invertible. The input augmented plant P_i has a state-space realization

$$P_i = \left[\begin{array}{c|cc} A & B_1 & \hat{B}_2 \\ \hline C_1 & D_{11} & \hat{D}_{12} \\ C_2 & D_{21} & 0 \end{array}\right] = \left[\begin{array}{c|ccc} A & B_1 & B_2' & B_2 \\ \hline C_1 & D_{11} & D_{12}' & D_{12} \\ C_2 & D_{21} & 0 & 0 \end{array}\right]. \tag{8.13}$$

Since $\hat{D}_{12}$ is invertible, P_i allows a dual chain-scattering representation $DCHAIN\,(P_i)$ which is given by

$$\begin{aligned} H &= DCHAIN\,(P_i) \\ &= \left[\begin{array}{c|cc} A - \hat{B}_2 \hat{D}_{12}^{-1} C_1 & \hat{B}_2 \hat{D}_{12}^{-1} & B_1 - \hat{B}_2 \hat{D}_{12}^{-1} D_{11} \\ \hline -\hat{D}_{12}^{-1} C_1 & \hat{D}_{12}^{-1} & -\hat{D}_{12}^{-1} D_{11} \\ C_2 & 0 & D_{21} \end{array}\right] \\ &:= \left[\begin{array}{c|c} A_H & B_H \\ \hline C_H & D_H \end{array}\right]. \end{aligned} \tag{8.14}$$

This representation is obtained by applying the identity (4.38) to (8.13).

Corresponding to the factorization (8.11) for $CHAIN\,(P_o)$, we obtain a similar factorization for $DCHAIN\,(P_i)$ in terms of partial plants

$$P_w := \begin{bmatrix} -I & P_{11} \\ 0 & P_{21} \end{bmatrix} = \left[\begin{array}{c|c} A & B_w \\ \hline C & D_w \end{array}\right], \tag{8.15}$$

$$\hat{P}_u := \begin{bmatrix} -\hat{P}_{12} & 0 \\ -\hat{P}_{22} & I \end{bmatrix} = \left[\begin{array}{c|c} A & \hat{B}_u \\ \hline C & \hat{D}_u \end{array}\right], \tag{8.16}$$

where

$$C = \begin{bmatrix} C_1 \\ C_2 \end{bmatrix}, \quad B_w = \begin{bmatrix} 0 & B_1 \end{bmatrix}, \quad D_w = \begin{bmatrix} -I & D_{11} \\ 0 & D_{21} \end{bmatrix}, \tag{8.17}$$

$$\hat{B}_u = \begin{bmatrix} -\hat{B}_2 & 0 \end{bmatrix} = \begin{bmatrix} -B_2' & -B_2 & 0 \end{bmatrix}, \tag{8.18}$$

$$\hat{D}_u = \begin{bmatrix} -\hat{D}_{12} & 0 \\ 0 & I \end{bmatrix} = \begin{bmatrix} -D_{12}' & -D_{12} & 0 \\ 0 & 0 & I \end{bmatrix}. \tag{8.19}$$

The partial plants (8.15) and (8.16) correspond to (4.39) and (4.40), respectively, and we have, from (4.33),

$$\begin{aligned} H &= DCHAIN\,(P_i) \\ &= DCHAIN\,(\begin{bmatrix} 0 & \hat{P}_{12} \\ I & \hat{P}_{22} \end{bmatrix}) \cdot DCHAIN\,(\begin{bmatrix} P_{11} & I \\ P_{21} & 0 \end{bmatrix}) \\ &= DCHAIN\,(\hat{P}_u \begin{bmatrix} 0 & -I \\ I & 0 \end{bmatrix}) \cdot DCHAIN\,(P_w \begin{bmatrix} 0 & -I \\ I & 0 \end{bmatrix}). \end{aligned}$$

It is worth noting that the second factor in this factorization does not depend on the augmentation (B_2', D_{12}').

Due to Lemma 4.3, we have another representation of $DCHAIN\,(P_i)$ as

$$H = \left[\begin{array}{c|c} A_H & B_H \\ \hline C_H & D_H \end{array}\right] = \left[\begin{array}{c|c} A - \hat{B}_u\hat{D}_u^{-1}C & B_w - \hat{B}_u\hat{D}_u^{-1}D_w \\ \hline \hat{D}_u^{-1}C & \hat{D}_u^{-1}D_w \end{array}\right]. \tag{8.20}$$

This representation implies that $H = DCHAIN\,(P_i)$ is obtained by applying an output insertion $-\hat{B}_u\hat{D}_u^{-1}$ and an output transformation with $\hat{D}_u^{-1}$ as its transformation matrix to a partial plant P_w that does not depend on the augmentation (B_2', D_{12}'). The block diagram of $H = DCHAIN\ (P_i)$ is given in Figure 8.2.

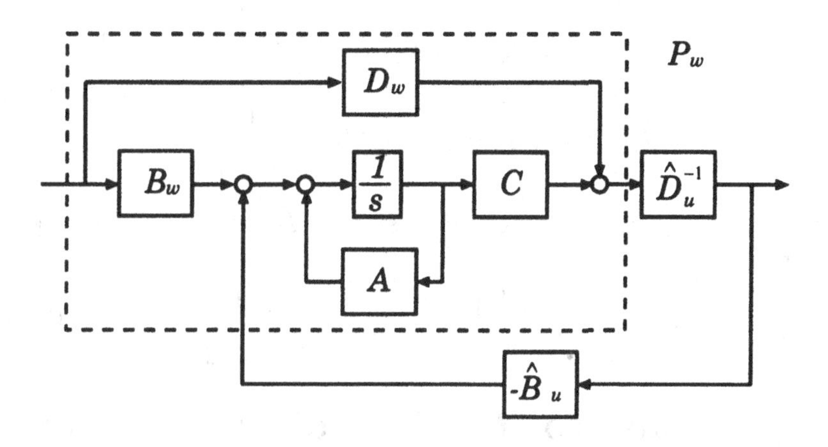

Figure 8.2 Block Diagram of $DCHAIN\ (P_i)$.

The dual chain-scattering representation (8.14) is in some sense an inverse of the chain-scattering representation (8.6), as was shown in (4.29), if both exist for a given plant. In case of augmentations, the relation between the two chain-scattering representations is given in the following result,

LEMMA 8.1 *G in (8.6) and H in (8.14) satisfy the following identities.*

$$A_G = A_H + B_H C_G, \tag{8.21a}$$

$$B_G = B_H D_G, \tag{8.21b}$$

$$C_H = -D_H C_G, \tag{8.21c}$$

$$D_H D_G = \begin{bmatrix} 0 & 0 \\ I_{p+q} & 0 \end{bmatrix}. \tag{8.21d}$$

The proof is straightforward based on the realizations (8.6) and (8.14). ∎

As a direct consequence of the preceding lemma, we have

$$H(s)G(s) = D_H D_G = \begin{bmatrix} 0 & 0 \\ I_{p+q} & 0 \end{bmatrix}, \tag{8.22}$$

which coincides with (7.68).

Before concluding the section, we note some relations satisfied by the realizations of partial plants (8.7), (8.8), (8.15), and (8.16):

$$\begin{aligned} D_w D_z = D_u D_y &= \begin{bmatrix} 0 & -D_{12} \\ D_{21} & 0 \end{bmatrix}, \\ D_y \hat{D}_y^{-1} = \begin{bmatrix} I & 0 \end{bmatrix}, \quad \hat{D}_u^{-1} D_u &= \begin{bmatrix} 0 \\ I \end{bmatrix}. \end{aligned} \tag{8.23}$$

where D_y and D_u are unaugmented versions of $\hat{D}_y$ and $\hat{D}_u$ given, respectively, by

$$D_y = \begin{bmatrix} 0 & I \\ D_{21} & 0 \end{bmatrix}, \quad D_u = \begin{bmatrix} -D_{12} & 0 \\ 0 & I \end{bmatrix}. \tag{8.24}$$

8.2 Solution to H^∞ Control Problem for Augmented Plants

In this section, we consider the normalized H^∞ control problem for augmented plants. We assume that the assumption ($A2_1$) holds in the output augmented plant P_o; that is,

$$\text{rank} \begin{bmatrix} A - j\omega I & B_1 \\ \hat{C}_2 & \hat{D}_{21} \end{bmatrix} = \text{rank} \begin{bmatrix} A - j\omega I & B_1 \\ C_2 & D_{21} \\ C_2' & D_{21}' \end{bmatrix} = n + r, \quad \forall \omega. \tag{8.25}$$

Since ($A2_1$) holds in the original plant, the condition (8.25) holds for almost all output augmentations (C_2', D_{21}'). An output augmentation (C_2', D_{21}') such that (8.25) holds and D_{21}' is a complement of D_{21}, is called *admissible*. Hence, if (C_2', D_{21}') is admissible, then A_G has no eigenvalues on the $j\omega$-axis. Therefore, $G = \mathit{CHAIN}\,(P)$ satisfies the condition of Theorem 7.4. Hence, the normalized H^∞ control problem is solvable iff $G = \mathit{CHAIN}\,(P)$ has a (J_{mr}, J_{pr})-lossless factorization (7.46) and the class of desired controllers is given by (7.47). Theorem 6.6 applied to G in (8.6) yields the following result.

THEOREM 8.2 *Under the assumptions (A1) and (A2), the normalized H^∞ control problem is solvable for each output augmented plant P_o given in (8.5) with admissible augmentation (C_2', D_{21}') iff the following conditions hold.*

(i) There exists a nonsingular matrix E_z such that

$$D_z^T J D_z = E_z^T J' E_z. \tag{8.26}$$

(ii) There exists a solution $X \geq 0$ of the algebraic Riccati equation

$$\begin{aligned} &XA_G + A_G^T X \\ &\quad -(C_G^T J D_G + X B_G)(D_G^T J D_G)^{-1}(D_G^T J C_G + B_G^T X) \\ &\qquad + C_G^T J C_G = 0 \end{aligned} \tag{8.27}$$

such that

$$\hat{A}_G := A_G + B_G F_G, \tag{8.28a}$$
$$F_G := -(D_G^T J D_G)^{-1}(D_G^T J C_G + B_G^T X) \tag{8.28b}$$

is stable.

(iii) There exists a solution $\bar{X} \geq 0$ to the algebraic Riccati equation

$$\bar{X} A_G^T + A_G \bar{X} + \bar{X} C_G^T J C_G \bar{X} = 0 \tag{8.29}$$

such that

$$\bar{A}_G := A_G + \bar{X} C_G^T J C_G \tag{8.30}$$

is stable.

(iv)

$$\sigma(X\bar{X}) < 1. \tag{8.31}$$

In that case, the class of controllers that solve the normalized H^∞ control problem is given by

$$K = HM(\Pi^{-1}; S), \tag{8.32}$$

where S is an arbitrary matrix in $\mathbf{BH}^\infty$ and

$$\Pi(s) = \left[\begin{array}{c|c} \bar{A}_G & -(B_G + \bar{X} C_G^T J D_G) \\ \hline E_z \hat{D}_y^{-1} F_G (I - \bar{X} X)^{-1} & E_z \hat{D}_y^{-1} \end{array}\right]. \tag{8.33}$$

In the above statements, we assumed $J = J_{mr}$, $J' = J_{vr}$.

Proof. The statements (i)~(iv) are almost the repetition of the conditions (i)~(iv) of Theorem 6.6 applied to $G = CHAIN\,(P_o)$, except Condition (i) where E satisfying (6.18) is given by

$$E = E_z \hat{D}_y^{-1}. \tag{8.34}$$

Indeed, since $D_G = D_z \hat{D}_y^{-1}$, E given by (8.34) satisfies (6.18) for $D = D_G$. ∎

Dualization of Theorem 8.2 is straightforward based on Theorem 7.5 and Theorem 6.12. An input augmentation (B_2', D_{12}') is said to be *admissible*, if

$$\begin{aligned}\text{rank}\begin{bmatrix} A - j\omega I & \hat{B}_2 \\ C_1 & \hat{D}_{12}\end{bmatrix} &= \text{rank}\begin{bmatrix} A - j\omega I & B_2' & B_2 \\ C_1 & D_{12}' & D_{12}\end{bmatrix} \\ &= n + m, \quad \forall\omega \end{aligned}\tag{8.35}$$

and D_{12}' is a complement of D_{12}.

THEOREM 8.3 *Under the assumptions (A1) and (A2), the normalized H^∞ control problem is solvable for each input augmented plant P_i given in (8.13) with admissible augmentation iff the following conditions hold.*

(i) There exists a non-singular matrix E_w satisfying

$$D_w J D_w^T = E_w J' E_w^T. \tag{8.36}$$

(ii) There exists a solution $Y \geq 0$ of the algebraic Riccati equation

$$\begin{aligned} &YA_H^T + A_H Y \\ &\quad + (B_H J D_H^T - Y C_H^T)(D_H J D_H^T)^{-1}(D_H J B_H^T - C_H Y) \\ &\qquad - B_H J B_H^T = 0 \end{aligned}\tag{8.37}$$

such that

$$\hat{A}_H := A_H + L_H C_H \tag{8.38a}$$

$$L_H := -(B_H J D_H^T - Y C_H^T)(D_H J D_H^T)^{-1} \tag{8.38b}$$

is stable.

(iii) There exists a solution $\bar{Y} \geq 0$ of the algebraic Riccati equation

$$\bar{Y}A_H + A_H^T\bar{Y} - \bar{Y}B_HJB_H^T\bar{Y} = 0 \tag{8.39}$$

such that

$$\bar{A}_H := A_H - B_HJB_H^T\bar{Y} \tag{8.40}$$

is stable.

(iv)

$$\sigma(Y\bar{Y}) < 1. \tag{8.41}$$

In that case, the class of controllers that solve the normalized H^∞ control problem for P_i is given by

$$K = DHM(\Omega^{-1}; S) \tag{8.42}$$

where S is an arbitrary matrix in $\mathbf{BH}^\infty$ *and*

$$\Omega(s) = \left[\begin{array}{c|c} \bar{A}_H & -(I - Y\bar{Y})^{-1}L_H\hat{D}_u^{-1}E_w \\ \hline C_H - D_HJB_H^T\bar{Y} & \hat{D}_u^{-1}E_w \end{array}\right]. \tag{8.43}$$

In the preceding statements, we assumed $J = J_{mr}$, $J' = J_{mq}$.

Theorems 8.2 and 8.3 give the complete solutions to the H^∞ control problems for the two block cases, namely, the case where either $r = q$ or $m = p$. In the case $r = q$, Assumption (A2$_3$) implies that D_{21} is invertible, which implies that $CHAIN\,(P)$ exists without output augmentation. We can just eliminate "$\wedge$" in the description (8.6) or (8.12) of G. Theorem 8.2 gives a complete solution to the normalized H^∞ control problem. The same remark applies to the case $m = p$ where D_{12} is invertible instead of D_{21}. In that case, Theorem 8.3 gives a complete solution to the normalized H^∞ control problem.

In order to compute the controller generator Π^{-1} from (8.33) and Ω^{-1} from (8.43), we note the following identities.

LEMMA 8.4 *In Theorem 8.2, the following relation holds.*

$$\bar{A}_G(I - \bar{X}X) + (B_G + \bar{X}C_G^TJD_G)F_G = (I - \bar{X}X)\hat{A}_G. \tag{8.44}$$

Dually, the following relation holds in Theorem 8.3:

$$(I - Y\bar{Y})\bar{A}_H + L_H(C_H - D_HJB_H^T\bar{Y}) = \hat{A}_H(I - Y\bar{Y}). \tag{8.45}$$

Proof. The Riccati equation (8.27) can be written as

$$A_G^T X + C_G^T J C_G = -XA_G - (C_G^T J D_G + XB_G)F_G.$$

Due to (8.29), we have

$$\begin{aligned}\bar{A}_G(I - \bar{X}X) &= (A_G + \bar{X}C_G^T J C_G)(I - \bar{X}X)\\ &= A_G + \bar{X}(C_G^T J C_G + A_G^T X)\\ &= A_G - \bar{X}(XA_G + (C_G^T J D_G + XB_G)F_G)\\ &= (I - \bar{X}X)(A_G + B_G F_G) - (B_G + \bar{X}C_G^T J D_G)F_G,\end{aligned}$$

which establishes the relation (8.44). The identity (8.45) can be proven by the dual argument. ∎

Based on Lemma 8.4, we can compute the controller generator $\Pi(s)^{-1}$ from (8.33). Due to the inversion rule (2.15) and (8.29), the A-matrix of $\Pi(s)^{-1}$ is calculated to be

$$\bar{A}_G + (B_G + \bar{X}C_G^T J D_G)F_G(I - \bar{X}X)^{-1} = (I - \bar{X}X)\hat{A}_G(I - \bar{X}X)^{-1}.$$

Here we have used the identity (8.44). Therefore, $\Pi(s)^{-1}$ is given by

$$\begin{aligned}\Pi(s)^{-1} &= \left[\begin{array}{c|c} (I - \bar{X}X)\hat{A}_G(I - \bar{X}X)^{-1} & (B_G + \bar{X}C_G^T J D_G)\\ \hline F_G(I - \bar{X}X)^{-1} & I\end{array}\right]\hat{D}_y E_z^{-1}\\ &= \left[\begin{array}{c|c} \hat{A}_G & (I - \bar{X}X)^{-1}(B_G + \bar{X}C_G^T J D_G)\\ \hline F_G & I\end{array}\right]\hat{D}_y E_z^{-1}. \qquad (8.46)\end{aligned}$$

Analogously, we obtain the dual controller generator $\Omega(s)^{-1}$ as

$$\Omega(s)^{-1} = E_w^{-1}\hat{D}_u\left[\begin{array}{c|c} \hat{A}_H & L_H\\ \hline (C_H - D_H J B_H^T \bar{Y})(I - Y\bar{Y})^{-1} & I\end{array}\right]. \qquad (8.47)$$

8.3 Maximum Augmentations

In the preceding section, we derived solvability conditions of the normalized H^∞ control problem for augmented plants. If the normalized H^∞ control problem is solvable for the original plant, it must be solved for augmented plants, because the solution to the original plant is also a special solution to the augmented plants which *does not* use the augmented input or output. Therefore, Theorems 8.2 and 8.3 give necessary conditions for the solvability of the H^∞ control problem for the original plant. It is natural to guess that a sufficient condition for the solvability must guarantee that the problem is solvable for *all* augmented plants with admissible augmentations. Actually, a sufficient condition will be derived by finding a special augmentation for which the (J, J')-lossless factorizations have lower triangular forms (7.72) as described in Theorem 7.7. The existence of such factorizations guarantees the existence of (J, J')-lossless factorizations for all augmented plants. We start with examining the conditions (i)~(iv) of Theorems 8.2 and 8.3 in detail.

The conditions (i) of the theorems do not depend on augmentations. Hence, they are necessary conditions for the solvability of the normalized H^∞ control problems for the original plant (8.1). Now we represent the conditions (i) of Theorems 8.2 and 8.3 in different ways.

LEMMA 8.5 *Condition (i) of Theorem 8.2 holds, that is, the identity (8.26) is satisfied for some nonsingular matrix E_z iff*

$$J - D_z(D_z^T J D_z)^{-1} D_z^T \geq 0. \tag{8.48}$$

Dually, Condition (i) of Theorem 8.3 holds, that is, the identity (8.36) holds for a nonsingular matrix E_w iff

$$J - D_w^T(D_w J D_w^T)^{-1} D_w \leq 0. \tag{8.49}$$

Proof. Let $R := J - D_z(D_z^T J D_z)^{-1} D_z^T$. Obviously, $RJD_z = 0$ and $RJR = R$. If (8.26) holds, we have

$$\begin{bmatrix} D_z^T \\ R \end{bmatrix} J \begin{bmatrix} D_z & R \end{bmatrix} = \begin{bmatrix} E_z^T J_{pr} E_z & 0 \\ 0 & R \end{bmatrix}.$$

Since $\mathrm{Ker}\begin{bmatrix} D_z & R \end{bmatrix} = \phi$ and $J = J_{mr}$, the inertia theorem implies that the matrix on the left-hand side has at most r negative eigenvalues. Since $E_z^T J_{pr} E_z$ has r negative eigenvalues, R has no negative eigenvalues.

Conversely, assume that (8.48) holds. Then we can find a matrix U of column full rank such that

$$J - D_z(D_z^T J D_z)^{-1} D_z^T = R = UU^T.$$

We can easily see that $\mathrm{Ker}U^T \subset \mathrm{Im}JD_z$. Therefore, since rank $D_z = r+p$ from (8.9), rank $U^T = (m+r) -$ rank $D_z = m - p$. Also, due to

$$UU^T = R = RJR = UU^T JUU^T,$$

we have $U^T JU = I_{m-p}$. It follows that

$$\begin{bmatrix} U^T \\ D_z^T \end{bmatrix} J \begin{bmatrix} U & D_z \end{bmatrix} = \begin{bmatrix} I_{m-p} & 0 \\ 0 & D_z^T J D_z \end{bmatrix}.$$

Again, the inertia theorem implies that $D_z^T J D_z$ has p positive eigenvalues and r negative eigenvalues. Thus, $D_z^T J_{mr} D_z$ must be congruent to J_{pr}. Thus, the first assertion has been established.

The second assertion can be proven by the dual argument. ∎

Lemma 8.5 implies that the inequalities (8.48) and (8.49) are necessary for the solvability of the H^∞ control problem for the normalized plant (8.1). In turn, they give explicit forms of the factorizations (8.26) and (8.36).

LEMMA 8.6 *Assume that the inequalities (8.48) and (8.49) hold. Then*

$$E_z := \begin{bmatrix} V_w D_y \\ U_w D_z \end{bmatrix} \tag{8.50}$$

satisfies (8.26), where U_w is a matrix with full row rank satisfying

$$J - D_w^T (D_w J D_w^T)^{-1} D_w = -U_w^T U_w \tag{8.51}$$

and V_w is a nonsingular matrix satisfying

$$D_u^T (D_w J D_w^T)^{-1} D_u = V_w^T J_{pq} V_w. \tag{8.52}$$

Dually,

$$E_w = \begin{bmatrix} D_w U_z & D_u V_z \end{bmatrix} \tag{8.53}$$

satisfies (8.36), where U_z is a matrix with full column rank satisfying

$$J - D_z (D_z^T J D_z)^{-1} D_z^T = U_z U_z^T \tag{8.54}$$

and V_z is a nonsingular matrix satisfying

$$D_y (D_z^T J D_z)^{-1} D_y^T = V_z J_{pq} V_z^T. \tag{8.55}$$

Note that $J = J_{mr}$ in the previous statements, and D_y and D_u are given by (8.24).

Proof. The existence of U_w satisfying (8.51) and U_z satisfying (8.54) are clear from the assumptions. We show the existence of a nonsingular V_w satisfying (8.52). Obviously, $D_z^T JU_z = 0$. Hence, $JU_z \in \mathrm{Ker} D_z^T$. From the forms of D_z in (8.9) and D_w in (8.17), we have

$$\mathrm{Ker} D_z^T = \begin{bmatrix} I \\ -D_{11}^T \end{bmatrix} \mathrm{Ker} D_{12}^T = D_w^T \mathrm{Ker} D_u^T.$$

Hence, $JU_z = D_w^T W$ for some $W \in \mathrm{Ker}\ D_u^T$. It follows that

$$\begin{bmatrix} W^T \\ D_u^T (D_w J D_w^T)^{-1} \end{bmatrix} D_w J D_w^T \begin{bmatrix} W & (D_w J D_w^T)^{-1} D_u \end{bmatrix}$$
$$= \begin{bmatrix} U_z^T J U_z & 0 \\ 0 & D_u^T (D_w J D_w^T)^{-1} D_u \end{bmatrix}.$$

From (8.54), it follows that $U_z U_z^T = U_z U_z^T J U_z U_z^T$. Since U_z is of full column rank, $U_z^T J U_z = I_{m-p}$. Since $D_w J D_w^T$ is congruent to J_{mq} from the assumption and Lemma 8.5, we conclude that $D_u^T (D_w J D_w^T)^{-1} D_u$ has p positive eigenvalues and q negative eigenvalues due to the inertia theorem. Thus, we have established the existence of a nonsingular matrix V_w satisfying (8.52). The identity (8.26) can be shown directly from

$$\begin{aligned} E_z^T J_{pr} E_z &= D_y^T V_w^T J_{pq} V_w D_y - D_z^T U_w^T U_w D_z \\ &= D_y^T D_u^T (D_w J D_w^T)^{-1} D_u D_y \\ &\qquad + D_z^T (J - D_w^T (D_w J D_w^T)^{-1} D_w) D_z \\ &= D_z^T J D_z, \end{aligned}$$

where we used the identity (8.23).

The proof of the dual assertion can be done analogously. ∎

Now we proceed to the analysis of the conditions (ii) of Theorems 8.1 and 8.2. Analogous to the conditions (i) of Theorems 8.1.and 8.2, they are also independent of the augmentations. To see this, we represent X and Y defined in these conditions in terms of the associated Hamiltonian matrices using the notations introduced in Section 6.4. According to the notations (6.69) and (6.71), we see that the Riccati equations (8.27) and (8.37) are represented, respectively, as

$$X = Ric\ f\left(\left[\begin{array}{c|c} A_G & B_G \\ \hline C_G & D_G \end{array}\right]\right), \quad Y = Ric\ f^{\sim}\left(\left[\begin{array}{c|c} A_H & B_H \\ \hline C_H & D_H \end{array}\right]\right). \tag{8.56}$$

$$f\left(\left[\begin{array}{c|c} A_G & B_G \\ \hline C_G & D_G \end{array}\right]\right) = \begin{bmatrix} A_G & 0 \\ -C_G^TJC_G & -A_G^T \end{bmatrix} + \begin{bmatrix} -B_G \\ C_G^TJD_G \end{bmatrix} (D_G^TJD_G)^{-1} \begin{bmatrix} D_G^TJC_G & B_G^T \end{bmatrix} \tag{8.57a}$$

$$f^\sim\left(\left[\begin{array}{c|c} A_H & B_H \\ \hline C_H & D_H \end{array}\right]\right) = \begin{bmatrix} A_H^T & 0 \\ -B_H^TJB_H & -A_H \end{bmatrix} - \begin{bmatrix} C_H^T \\ B_HJD_H^T \end{bmatrix} (D_HJD_H^T)^{-1} \begin{bmatrix} D_HJB_H^T & -C_H \end{bmatrix}. \tag{8.57b}$$

Due to (8.12) and (6.77),

$$f\left(\left[\begin{array}{c|c} A_G & B_G \\ \hline C_G & D_G \end{array}\right]\right) = f\left(\left[\begin{array}{c|c} A+BF_y & BU_y \\ \hline C_z+D_zF_y & D_zU_y \end{array}\right]\right) = f\left(\left[\begin{array}{c|c} A & B \\ \hline C_z & D_z \end{array}\right]\right), \tag{8.58}$$

where we put $F_y = -\hat{D}_y^{-1}C_y$, $U_y = \hat{D}_y^{-1}$. Similarly, from (8.20) and (6.78), it follows that

$$f^\sim\left(\left[\begin{array}{c|c} A_H & B_H \\ \hline C_H & D_H \end{array}\right]\right) = f^\sim\left(\left[\begin{array}{c|c} A-\hat{B}_u\hat{D}_u^{-1}C & B_w-\hat{B}_u\hat{D}_u^{-1}D_w \\ \hline \hat{D}_u^{-1}C & \hat{D}_u^{-1}D_w \end{array}\right]\right) = f^\sim\left(\left[\begin{array}{c|c} A & B_w \\ \hline C & D_w \end{array}\right]\right). \tag{8.59}$$

The relation (8.58) implies that we can replace G by a much simpler "partial plant" P_z given in (8.7) to characterize X; that is, X is a solution to the algebraic Riccati equation

$$XA + A^TX - (C_z^TJD_z + XB)(D_z^TJD_z)^{-1}(D_z^TJC_z + B^TX) + C_z^TJC_z = 0 \tag{8.60}$$

which stabilizes

$$\hat{A}_G = A + BF, \tag{8.61}$$

$$F := -(D_z^TJD_z)^{-1}(D_z^TJC_z + B^TX). \tag{8.62}$$

The equivalence of (8.27) and (8.60) can also be shown by direct manipulations. Equation (8.27) is written as

$$X(A_G + B_G F_G) + (A_G + B_G F_G)^T X \\ + (C_G + D_G F_G)^T J (C_G + D_G F_G) = 0, \qquad (8.63)$$

where F_G is given by (8.28b). Routine calculations using (8.12) and (8.28b) yield

$$\begin{aligned} F_G &= -((D_z \hat{D}_y^{-1})^T J D_z \hat{D}_y^{-1})^{-1} (\hat{D}_y^{-T} D_z^T J (C_z - D_z \hat{D}_y^{-1} \hat{C}_y) + \hat{D}_y^{-T} B^T X^T) \\ &= \hat{D}_y F + \hat{C}_y. \end{aligned}$$

Using this relation, we have

$$A_G + B_G F_G = A + BF, \qquad (8.64)$$

$$C_G + D_G F_G = C_z + D_z F. \qquad (8.65)$$

The identity (8.64) establishes (8.61). Equation (8.63) is now written as

$$X(A + BF) + (A + BF)^T X + (C_z + D_z F)^T J (C_z + D_z F) = 0. \qquad (8.66)$$

Substitution of (8.62) in this equation yields (8.60).

Dually, the relation (8.59) implies that we can replace H by a much simpler "partial plant" P_w given in (8.15) to characterize Y; that is, Y is a solution to the algebraic Riccati equation

$$YA^T + AY + (B_w J D_w^T - YC^T)(D_w J D_w^T)^{-1}(D_w J B_w^T - CY) \\ - B_w J B_w^T = 0, \qquad (8.67)$$

which stabilizes

$$\hat{A}_H = A + LC, \qquad (8.68)$$

$$L := -(B_w J D_w^T - YC^T)(D_w J D_w^T)^{-1}. \qquad (8.69)$$

Straightforward computation using (8.20), (8.38b), and (8.69) shows

$$L_H = \hat{B}_u + L\hat{D}_u.$$

The equivalence of Equations (8.37) and (8.67) can be proven directly based on the identities

$$A_H + L_H C_H = A + LC, \qquad (8.70)$$

$$B_H + L_H D_H = B_w + L D_w. \qquad (8.71)$$

An alternative representation of (8.67) is given as

$$Y(A+LC)^T + (A+LC)Y - (B_w + LD_w)J(B_w + LD_w)^T = 0 \quad (8.72)$$

This representation corresponds to (8.66). Thus, we have established the following result.

LEMMA 8.7 *The solution X of the Riccati equation (8.27) is independent of the augmentation (C_2', D_{21}') and is characterized by*

$$X = Ric\ f\left(\left[\begin{array}{c|c} A & B \\ \hline C_z & D_z \end{array}\right]\right), \quad (8.73)$$

and the solution Y of the Riccati equation (8.37) is independent of the augmentation (B_2', D_{12}') and is characterized by

$$Y = Ric\ f^{\sim}\left(\left[\begin{array}{c|c} A & B_w \\ \hline C & D_w \end{array}\right]\right). \quad (8.74)$$

Now we have shown that the first two conditions of Theorems 8.1 and 8.2 do not depend on the augmentations.

The conditions (iii) in Theorems 8.1 and 8.2 do depend on the augmentations because the solutions $\bar{X}$ and $\bar{Y}$ of the Riccati equations (8.29) and (8.39), respectively, depend on the augmentations. In terms of the notations of Section 6.4, they can be represented as

$$\bar{X} = Ric\ g\left(\left[\begin{array}{c|c} A_G & B_G \\ \hline C_G & D_G \end{array}\right]\right) = Ric\begin{bmatrix} A_G^T & C_G^T J C_G \\ 0 & -A_G \end{bmatrix}, \quad (8.75a)$$

$$\bar{Y} = Ric\ g^{\sim}\left(\left[\begin{array}{c|c} A_H & B_H \\ \hline C_H & D_H \end{array}\right]\right) = Ric\begin{bmatrix} A_H & -B_H J B_H^T \\ 0 & -A_H^T \end{bmatrix}. \quad (8.75b)$$

The solution $\bar{X}$ depends on output augmentation (C_2', D_{21}'), whereas $\bar{Y}$ depends on input augmentation (B_2', D_{12}'). Now we show that the solution $X(Y)$ of the Riccati equation (8.27)(Riccati equation (8.37)) which is independent of the augmentations can be regarded as a sort of maximum solution of the Riccati equation (8.39)(the Riccati equation (8.29)) with respect to the input(output) augmentations.

THEOREM 8.8 *Assume that the normalized H^∞ control problem is solvable for the original plant (8.1) under the assumptions (A1) and (A2). Then, for each admissible input augmentation, the inequality*

$$\bar{Y} \leq X \tag{8.76}$$

is satisfied. The equality holds iff the input augmentation (B_2', D_{12}') is chosen such that

$$C_z + D_z F + J B_H^T X = 0, \tag{8.77}$$

where F is given by (8.62).

Dually, for each admissible output augmentation,

$$\bar{X} \leq Y \tag{8.78}$$

is satisfied. The equality holds iff the output augmentation (C_2', D_{21}') is chosen such that

$$B_w + L D_w + Y C_G^T J = 0, \tag{8.79}$$

where L is given by (8.69).

Proof. Due to (8.57a), (8.21a), and (8.21b),

$$f\left(\left[\begin{array}{c|c} A_G & B_G \\ \hline C_G & D_G \end{array}\right]\right) = \begin{bmatrix} A_H + B_H C_G & 0 \\ -C_G^T J C_G & -A_H^T - C_G^T B_H^T \end{bmatrix}$$
$$+ \begin{bmatrix} -B_H \\ C_G^T J \end{bmatrix} D_G (D_G^T J D_G)^{-1} D_G \begin{bmatrix} J C_G & B_H^T \end{bmatrix}.$$

Using (8.75b), we can show the identity

$$f\left(\left[\begin{array}{c|c} A_G & B_G \\ \hline C_G & D_G \end{array}\right]\right) - g^{\sim}\left(\left[\begin{array}{c|c} A_H & B_H \\ \hline C_H & D_H \end{array}\right]\right)$$
$$= \begin{bmatrix} B_H \\ -C_G^T J \end{bmatrix} (J - D_G (D_G^T J D_G)^{-1} D_G^T) \begin{bmatrix} J C_G & B_H^T \end{bmatrix}.$$

Since $D_G = D_z \hat{D}_y^{-1}$, we have

$$J - D_G (D_G^T J D_G)^{-1} D_G = J - D_z (D_z^T J D_z)^{-1} D_z^T.$$

Therefore, due to Lemma 8.5, we have

$$\left(f\left(\left[\begin{array}{c|c} A_G & B_G \\ \hline C_G & D_G \end{array}\right]\right) - g^{\sim}\left(\left[\begin{array}{c|c} A_H & B_H \\ \hline C_H & D_H \end{array}\right]\right)\right) \begin{bmatrix} 0 & -I \\ I & 0 \end{bmatrix} \geq 0.$$

Let

$$H_1 := f\left(\left[\begin{array}{c|c} A_G & B_G \\ \hline C_G & D_G \end{array}\right]\right) = f\left(\left[\begin{array}{c|c} A & B \\ \hline C_z & D_z \end{array}\right]\right),$$

$$H_2 := g^{\sim}\left(\left[\begin{array}{c|c} A_H & B_H \\ \hline C_H & D_H \end{array}\right]\right).$$

From the preceding inequality, these two Hamiltonian matrices satisfy the conditions of Lemma 3.7. Hence, the inequality (8.74) is a direct consequence of Lemma 3.7.

Due to Lemma 3.7, the equality holds iff

$$(J - D_G(D_G^T J D_G)^{-1} D_G^T)(JC_G + B_H^T X) = 0.$$

Noting the relation (8.21b), we easily see that this relation can be written as

$$C_G + JB_H^T X + D_G F_G = 0,$$

where F_G is given by (8.28a). Due to the identity (8.65), this relation identical to (8.77).

The dual assertion can be proven based on the relation

$$f^{\sim}\left(\left[\begin{array}{c|c} A_H & B_H \\ \hline C_H & D_H \end{array}\right]\right) - g\left(\left[\begin{array}{c|c} A_G & B_G \\ \hline C_G & D_G \end{array}\right]\right)$$
$$\begin{bmatrix} -C_G^T \\ B_H J \end{bmatrix} (J - D_H^T (D_H J D_H^T)^{-1} D_H) \begin{bmatrix} JB_H & C_G \end{bmatrix},$$

which again implies that

$$\left(f^{\sim}\left(\left[\begin{array}{c|c} A_H & B_H \\ \hline C_H & D_H \end{array}\right]\right) - g\left(\left[\begin{array}{c|c} A_G & B_G \\ \hline C_G & D_G \end{array}\right]\right)\right) \begin{bmatrix} 0 & I \\ -I & 0 \end{bmatrix} \geq 0$$

according to (8.49). The rest of the proof is left to the reader. ∎

Each input augmentation that satisfies the equality in (8.76) is called a *maximum input augmentation*, and each input augmentation that satisfies the equality in (8.78) is called a *maximum output augmentation*. The following result gives more explicit characterizations of the maximum augmentations.

LEMMA 8.9 *An output augmentation (C_2', D_{21}') is maximum iff*

$$D_{21}'(LD_c + B_1)^T + C_2'Y = 0. \tag{8.80}$$

where

$$D_c := \begin{bmatrix} D_{11} \\ D_{21} \end{bmatrix} = D_w \begin{bmatrix} 0 \\ I \end{bmatrix}. \tag{8.81}$$

Dually, an input augmentation (B_2', D_{12}') is maximum iff

$$(D_r F + C_1)^T D_{12}' + XB_2' = 0, \tag{8.82}$$

where

$$D_r := \begin{bmatrix} D_{11} & D_{12} \end{bmatrix} = \begin{bmatrix} I & 0 \end{bmatrix} D_z. \tag{8.83}$$

Proof. Due to Theorem 8.8, it is sufficient to prove that

$$M := B_w + LD_w + YC_G^T J = 0$$

iff (8.80) holds. From (8.17) and (8.6), it follows that

$$D_w C_G = \begin{bmatrix} -I & D_{11} \\ 0 & D_{21} \end{bmatrix} \begin{bmatrix} C_1 - D_{11}\hat{D}_{21}^{-1}\hat{C}_2 \\ -\hat{D}_{21}^{-1}\hat{C}_2 \end{bmatrix} = - \begin{bmatrix} C_1 \\ C_2 \end{bmatrix} = -C.$$

According to the definition (8.69) of L, we have

$$D_w J(B_w + LD_w)^T + D_w C_G Y = D_w J(B_w + LD_w)^T - CY = 0,$$

which implies that $D_w J M^T = 0$. Since $\begin{bmatrix} -I & D_{11} \\ 0 & \hat{D}_{21} \end{bmatrix} = \begin{bmatrix} D_w \\ \begin{bmatrix} 0 & D_{21}' \end{bmatrix} \end{bmatrix}$ is invertible, the identity (8.79) holds iff

$$\begin{bmatrix} 0 & D_{21}' \end{bmatrix} M^T = \begin{bmatrix} 0 & D_{21}' \end{bmatrix} J(LD_w + B_w)^T - C_2'Y = 0,$$

which is equal to (8.80). The condition (8.82) is similarly proven. ∎

Before concluding the section, we remark on an interesting property of the maximum augmentations.

LEMMA 8.10 *If an input augmentation (B_2', D_{12}') is maximum, then*

$$\bar{A}_H = \hat{A}_G. \tag{8.84}$$

Also, if an output augmentation (C_2', D_{21}') is maximum, then

$$\bar{A}_G = \hat{A}_H. \tag{8.85}$$

Proof. Due to (8.77) and $\bar{Y} = X$, we have

$$\begin{aligned}\bar{A}_H &= A_H - B_H J B_H^T \bar{Y} = A_H - B_H J B_H^T X \\ &= A_H + B_H(C_z + D_z F).\end{aligned}$$

From (8.65), (8.21a), and (8.21b), the identity (8.84) follows immediately. The identity (8.85) can be shown similarly. ∎

8.4 State-Space Solutions

Based on the results of the preceding sections, we can now state the solution to the normalized H^∞ control problem. The solution is actually obtained by solving the problem for a maximally augmented plant.

THEOREM 8.11 *Under the assumptions (A1) and (A2), the normalized H^∞ control problem is solvable for the plant (8.1) iff*

(i) D_z given by (8.9) satisfies the inequality

$$J - D_z(D_z^T J D_z)^{-1} D_z^T \geq 0.$$

(ii) D_w given by (8.17) satisfies the inequality

$$J - D_w^T(D_w J D_w^T)^{-1} D_w \leq 0.$$

(iii) There exists a solution $X \geq 0$ of the algebraic Riccati equation

$$\begin{aligned}XA + A^T X - (C_1^T D_r + XB)(D_z^T J D_z)^{-1}(D_r^T C_1 + B^T X)& \\ + C_1^T C_1 = 0,& \qquad (8.86)\\ D_r := \begin{bmatrix} D_{11} & D_{12} \end{bmatrix},& \qquad (8.87)\end{aligned}$$

which stabilizes

$$\begin{aligned}\hat{A}_G &:= A + BF, \\ F &= -(D_z^T J D_z)^{-1}(D_z^T J C_z + B^T X),\end{aligned}$$

or equivalently, there exists

$$X = Ric\, f\left(\left[\begin{array}{c|c} A & B \\ \hline C_z & D_z \end{array}\right]\right) \geq 0. \qquad (8.88)$$

(iv) There exists a solution $Y \geq 0$ of the algebraic Riccati equation

$$YA^T + AY + (B_1D_c^T + YC^T)(D_wJD_w^T)^{-1}(D_cB_1^T + CY) + B_1B_1^T = 0, \tag{8.89}$$

$$D_c := \begin{bmatrix} D_{11} \\ D_{21} \end{bmatrix}, \tag{8.90}$$

which stabilizes

$$\hat{A}_H := A + LC,$$
$$L = -(B_1D_c^T + YC^T)(D_wJD_w^T)^{-1},$$

or, equivalently, there exists

$$Y = Ric\ f^{\sim}\left(\left[\begin{array}{c|c} A & B_w \\ \hline C & D_w \end{array}\right]\right) \geq 0. \tag{8.91}$$

(v)

$$\sigma(XY) < 1. \tag{8.92}$$

In that case, a desired controller is given by

$$K = HM\ (\Pi_{11}^{-1}; S), \tag{8.93}$$

where S is any matrix in $\mathbf{BH}^\infty$,

$$\Pi_{11} = V_w \left[\begin{array}{c|c} A + LC & B_u + LD_u \\ \hline (D_yF + C_y)(I - YX)^{-1} & I \end{array}\right], \tag{8.94}$$

and V_w is given by (8.52).

Proof. *Necessity.* If the normalized H^∞ control problem is solvable for the original plant (8.1), it must be solvable for augmented plants with admissible augmentations. Therefore, Theorems 8.2 and 8.3 must hold. The conditions (i) and (ii) are equivalent to Condition (i) of Theorem 8.2 and Condition (i) of Theorem 8.3, respectively, due to Lemma 8.5. Hence, they are necessary. Conditions (iii) and (iv) are equivalent to Condition (ii) of Theorem 8.2 and Condition (ii) of Theorem 8.3, respectively, due to Lemma 8.7. Note that the Riccati equations (8.60) and (8.67) are further simplified to (8.86) and (8.89), respectively, by noting that $C_z^T J D_z = C_1^T D_r$ due to (8.9) and $B_w J D_w^T = -B_1 D_c^T$ due to (8.17). If we consider a maximum output augmentation, the solution $\bar{X}$ of (8.29) satisfies $\bar{X} = Y$. Hence, Condition (v) follows immediately from Condition (iv) of Theorem 8.2.

Sufficiency. Consider an output augmented plant (8.5) with maximum augmentation. If the conditions (i)~(v) hold, Theorem 8.2 obviously holds with $\bar{X} = Y$. Therefore, in view of Theorem 7.7, it is sufficient to show that the unimodular factor $\Pi(s)$ given by (8.33) is of lower triangular form (7.72). Since the relation (8.79) holds for each maximum input augmentation, we have

$$\begin{aligned} B_G + \bar{X} C_G^T J D_G &= (B + Y C_G^T J D_z)\hat{D}_y^{-1} \\ &= (B - (B_w + L D_w) D_z)\hat{D}_y^{-1} \\ &= \left(\begin{bmatrix} 0 & B_2 \end{bmatrix} - L D_w D_z\right)\hat{D}_y^{-1}. \end{aligned}$$

From (8.23) and (8.24), it follows that

$$B_G + \bar{X} C_G^T J D_G = \begin{bmatrix} -(B_u + L D_u) & 0 \end{bmatrix}.$$

Also, from (8.28a) and (8.12), it follows that $\hat{D}_y^{-1} F_G = F + \hat{D}_y^{-1}\hat{C}_y$. Let us take (8.50) as a factor. Then,

$$E_z \hat{D}_y^{-1} F_G = \begin{bmatrix} V_w(D_y F + C_y) \\ U_w(D_z F + D_z \hat{D}_y^{-1}\hat{C}_y) \end{bmatrix},$$

$$E_z \hat{D}_y^{-1} = \begin{bmatrix} V_w & 0 \\ E_{21} & E_{22} \end{bmatrix}, \quad \begin{bmatrix} E_{21} & E_{22} \end{bmatrix} = U_w D_z \hat{D}_y^{-1}.$$

Therefore, we conclude that $\Pi(s)$ given by (8.33) is given by

$$\Pi(s) = \begin{bmatrix} \Pi_{11}(s) & 0 \\ \Pi_{21}(s) & \Pi_{22}(s) \end{bmatrix},$$

where $\Pi_{11}(s)$ is given by (8.94) and

$$\begin{aligned} &\begin{bmatrix} \Pi_{21}(s) & \Pi_{22}(s) \end{bmatrix} \\ = \quad & U_w D_z \left[\begin{array}{c|c} A + LC & B_u + LD_u \;\; 0 \\ \hline (F + \hat{D}_y^{-1}\hat{C}_y)(I - \bar{X}X)^{-1} & \hat{D}_y^{-1} \end{array}\right]. \end{aligned}$$

Thus, we have established the theorem. ∎

The dual representation of the controller (8.93) is given by

$$K = DHM\,(\Omega_{22}^{-1}; S), \tag{8.95}$$

where S is any matrix in $\mathbf{BH}^\infty$,

$$\Omega_{22} = \left[\begin{array}{c|c} A+BF & -(I-YX)^{-1}(B_u+LD_u) \\ \hline C_y+D_yF & I \end{array}\right] V_z, \tag{8.96}$$

where V_z is a nonsingular matrix satisfying (8.55). The controller generator Π_{11}^{-1} can be explicitly computed based on the representation (8.46) as

$$\Pi_{11}^{-1} = \left[\begin{array}{c|c} A+BF & -(I-YX)^{-1}(B_u+LD_u) \\ \hline C_y+D_yF & I \end{array}\right] V_w^{-1}. \tag{8.97}$$

We can derive the unnormalized version of Theorem 8.8 using the scaled plant (7.36a); that is,

$$P_\gamma := \begin{bmatrix} \gamma^{-1}P_{11} & P_{12} \\ \gamma^{-1}P_{21} & P_{22} \end{bmatrix} = \left[\begin{array}{c|cc} A & B_1 & \gamma B_2 \\ \hline \gamma^{-1}C_1 & \gamma^{-1}D_{11} & D_{12} \\ \gamma^{-1}C_2 & \gamma^{-1}D_{21} & 0 \end{array}\right]. \tag{8.98}$$

Let

$$\Sigma_\gamma := \begin{bmatrix} \gamma^{-1}I & 0 \\ 0 & I \end{bmatrix}, \tag{8.99}$$

$$J_\gamma := \Sigma_\gamma J \Sigma_\gamma = \begin{bmatrix} \gamma^{-2}I & 0 \\ 0 & -I \end{bmatrix}. \tag{8.100}$$

To get the solvability condition for the plant P_γ, the following replacements in Theorem 8.11 are in order.

$$D_z \rightarrow \begin{bmatrix} \gamma^{-1}D_{11} & D_{12} \\ I & 0 \end{bmatrix} = \gamma\Sigma_\gamma D_z \Sigma_\gamma. \tag{8.101}$$

$$D_w \rightarrow \begin{bmatrix} -I & \gamma^{-1}D_{11} \\ 0 & \gamma^{-1}D_{21} \end{bmatrix} = \gamma^{-1}D_w\Sigma_\gamma^{-1}. \tag{8.102}$$

$$C \rightarrow \begin{bmatrix} \gamma^{-1}C_1 \\ \gamma^{-1}C_2 \end{bmatrix} = \gamma^{-1}C. \tag{8.103}$$

$$B \rightarrow [B_1 \ \gamma B_2] = B\Sigma_\gamma \gamma. \tag{8.104}$$

THEOREM 8.12 *The unnormalized H^∞ control problem (7.29) is solvable iff*

(i) D_z given by (8.101) satisfies the inequality

$$J_\gamma^{-1} - D_z(D_z^T J_\gamma D_z)^{-1} D_z^T \geq 0. \tag{8.105}$$

(ii) D_w given by (8.102) satisfies the inequality

$$J_\gamma - D_w^T(D_w J_\gamma^{-1} D_w^T)^{-1} D_w \leq 0. \tag{8.106}$$

(iii) There exists a solution $X \geq 0$ of the algebraic Riccati equation

$$XA + A^T X - \frac{1}{\gamma^2}(C_1^T D_r + XB)(D_z^T J_\gamma D_z)^{-1}(D_r^T C_1 + B^T X) + C_1^T C_1 = 0 \tag{8.107}$$

which stabilizes $\hat{A}_G = A + BF$ where

$$F = -\frac{1}{\gamma^2}(D_z^T J_\gamma D_z)^{-1}(D_r^T C_1 + B^T X). \tag{8.108}$$

(iv) There exists a solution $Y \geq 0$ of the Riccati equation

$$YA^T + AY - (B_1 D_c^T + YC^T)(D_w J_\gamma^{-1} D_w^T)^{-1}(D_c B_1^T + CY) + B_1 B_1^T = 0, \tag{8.109}$$

which stabilizes $\hat{A}_H = A + LC$ with

$$L = -(B_1 D_c^T + YC^T)(D_w J_\gamma^{-1} D_w^T)^{-1}. \tag{8.110}$$

(v)

$$\sigma(XY) < \gamma^2. \tag{8.111}$$

In that case, a desired controller is given by (8.93) where S is any matrix in $\mathbf{BH}^\infty$,

$$\Pi_{11} = V_w \left[\begin{array}{c|c} A + LC & B_u + LD_u \\ \hline (D_y F + C_y)(I - \gamma^{-2} Y X) & I \end{array} \right] \tag{8.112}$$

and V_w is given by

$$D_u^T (\gamma^{-2} D_w J_\gamma^{-1} D_w^T)^{-1} D_u = V_w^T J V_w. \tag{8.113}$$

Proof. The inequality (8.105) is easily derived from (i) of Theorem 8.11 and (8.101). Also, the inequality (8.106) is similarly derived from (ii) of Theorem 8.11 and (8.102). The substitution of (8.101)~(8.104) in (8.86) yields

$$\begin{aligned} & XA + A^T X \\ & \quad -(\gamma^{-1} C_1^T D_r + \gamma X B)(\gamma^2 D_z^T J_\gamma D_z)^{-1} (\gamma^{-1} D_r^T C_1 + \gamma B^T X) \\ & \qquad + \gamma^{-2} C_1^T C_1 = 0. \end{aligned} \tag{8.114}$$

Multiplication of (8.114) by γ^2 yields

$$\begin{aligned} \gamma^2 XA + A^T X \gamma^2 - \frac{1}{\gamma^2} (C_1^T D_r + \gamma^2 X B)(D_z^T J_\gamma D_z)^{-1} (D_r^T C_1 + B^T X \gamma^2) \\ + C_1^T C_1 = 0. \end{aligned}$$

Taking $\gamma^2 X$ as a new X yields (8.107). The replacement (8.104) yields

$$\begin{aligned} \hat{A}_G &= A + BF = A + B\gamma \Sigma_\gamma F \\ \gamma \Sigma_\gamma F &= -\frac{1}{\gamma^2} (D_z^T J_\gamma D_z)^{-1} (D_r^T C_1 + B^T X \gamma^2). \end{aligned}$$

Therefore, the assertion (iii) has been proven. Direct computation of (8.90) with the replacements (8.101) to (8.104) yields (8.109). Also, we have

$$\begin{aligned} \hat{A}_H &= A + \gamma^{-1} LC, \\ \gamma^{-1} L &= -\gamma^{-1} (B_1 D_c^T \gamma^{-1} + Y \gamma^{-1} C^T)(\gamma^{-2} D_w J_\gamma D_w^T)^{-1} \\ &= -(B_1 D_c^T + Y C^T)(D_w J_\gamma D_w^T)^{-1}. \end{aligned}$$

Thus, the statement (iii) has been proven. The replacement $X \to \gamma^2 X$ in (iv) of Theorem 8.11 yields (8.111). It is straightforward to derive (8.112) from (8.94). ∎

Similar computation to derive (8.97) yields the controller generator

$$\Pi_{11}^{-1} = \left[\begin{array}{c|c} A+BF & -(I-\gamma^{-2}XY)^{-1}(B_u+LD_u) \\ \hline C_y+D_yF & I \end{array}\right] V_w^{-1}.$$

8.5 Some Special Cases

In this section, we show that the solvability condition given by Theorem 8.12 can be simplified for several special cases of practical importance. Simplification of the solvability condition gives a deeper insight into the structure of H^∞control.

(i) The case where D_{12} is invertible.
This case corresponds to one of the so-called two-block cases. In this case, D_z is invertible. Hence, Condition (i) of Theorem 8.12 holds with equality. The Riccati equation (8.107) is written in this case as

$$X(A-B_2D_{12}^{-1}C_1)+(A-B_2D_{12}^{-1}C_1)X-\frac{1}{\gamma^2}XB(D_z^TJ_\gamma D_z)^{-1}B^TX=0 \tag{8.115}$$

and F in (8.108) is given by

$$F=\frac{1}{\gamma^2}\left[\begin{array}{c} -I \\ D_{12}^TD_{11} \end{array}\right](B_1^T-D_{11}D_{12}^{-T}B_2^T)X + \left[\begin{array}{c} 0 \\ D_{12}^{-1}(D_{12}^{-T}B_2^TX-C_1) \end{array}\right]. \tag{8.116}$$

The Riccati equation (8.115) is degenerate in the sense that it lacks the constant term. If $A-B_2D_{12}^{-1}C_1$ is stable, then $X=0$ is a solution of (8.115) which stabilizes $\hat{A}_G=A+BF=A-B_2D_{12}^{-1}C_1$. Therefore, Condition (iii) of Theorem 8.12 is unnecessary. Also, Condition (v) holds automatically for any γ.

(ii) The case where D_{21} is invertible.
This corresponds to another two-block case. Since D_w is invertible in this case, Condition (ii) of Theorem 8.12 holds with equality. The Riccati equation (8.109) is written in this case as

$$Y(A-B_1D_{21}^{-1}C_2)^T+(A-B_1D_{21}^{-1}C_2)Y -\frac{1}{\gamma^2}YC^T(D_wJ_\gamma^{-1}D_w^T)^{-1}CY=0 \tag{8.117}$$

and L in (8.110) is given by

$$L = \frac{1}{\gamma^2} Y(C_1 - D_{11}D_{21}^{-1}C_2)^T \begin{bmatrix} -I & D_{11}D_{21}^{-1} \end{bmatrix} + \begin{bmatrix} 0 & (YC_2^T D_{21}^{-T} - B_1)D_{21}^{-1} \end{bmatrix}. \quad (8.118)$$

If $A - B_1D_{21}^{-1}C_2$ is stable, $Y = 0$ is a solution of (8.117) that stabilizes $\hat{A}_H = A + LC = A - B_1D_{21}^{-1}C_2$. Therefore, Condition (iv) of Theorem 8.12 always holds with $Y = 0$. Condition (v) is satisfied automatically for any γ. We summarize the reasoning.

COROLLARY 8.13 *If the assumptions of Theorem 8.2 hold and $D_{12}(D_{21})$ is invertible, then the H^∞control problem is solvable iff the conditions (ii)~(v) ((i), (iii)~(v)) of Theorem 8.12 hold with (8.107) and (8.108) being replaced by (8.115) and (8.116), respectively (with (8.109) and (8.110) being replaced by (8.117) and (8.118)). If further, $A - B_2D_{12}^{-1}C_1(A - B_1D_{21}^{-1}C_2)$ is stable, then the H^∞control problem is solvable iff the conditions (i) and (iii) ((ii) and (iv)) of Theorem 8.12 hold.*

Remark: The latter part of this corollary implies that the stability of $A - B_2D_{12}^{-1}C_1$ or $A - B_1D_{21}^{-1}C_2$ is a crucial factor of the solvability conditions. Obviously, $A - B_2D_{12}^{-1}C_1$ is stable iff P_{12}^{-1} is stable and $A - B_1D_{21}^{-1}C_2$ is stable iff P_{21}^{-1} is stable.

(iii) The case $D_{11} = 0$.
In this case,

$$D_z = \begin{bmatrix} 0 & D_{12} \\ I & 0 \end{bmatrix}, \quad D_w = \begin{bmatrix} -I & 0 \\ 0 & D_{21} \end{bmatrix}.$$

It follows that

$$J_\gamma^{-1} - D_z(D_z^T J_\gamma D_z)^{-1} D_z^T = \gamma^2 \begin{bmatrix} I - D_{12}R_z^{-1}D_{12}^T & 0 \\ 0 & 0 \end{bmatrix},$$

$$J_\gamma - D_w^T(D_w J_\gamma^{-1} D_w)^{-1} D_w = \begin{bmatrix} 0 & 0 \\ 0 & D_{21}^T R_w D_{21} - I \end{bmatrix},$$

where

$$R_z := D_{12}^T D_{12}, \quad R_w := D_{21}D_{21}^T. \quad (8.119)$$

Since $D_{12}R_z^{-1}D_{12}^T \le I$ and $D_{21}^T R_w^{-1} D_{21} \le I$, the inequalities (8.105) and (8.106) are both satisfied in this case. The Riccati equation (8.107) becomes in this case

$$XA + A^TX + \frac{1}{\gamma^2}XB_1B_1^TX - (C_1^TD_{12} + XB_2)R_z^{-1}(D_{12}^TC_1 + B_2^TX) + C_1^TC_1 = 0 \tag{8.120}$$

and F in (8.108) is given by

$$F = \begin{bmatrix} \frac{1}{\gamma^2}B_1^TX \\ -R_z^{-1}(D_{12}^TC_1 + B_2^TX) \end{bmatrix} := \begin{bmatrix} F_w \\ F_u \end{bmatrix}. \tag{8.121}$$

The Riccati equation (8.109) becomes in this case

$$YA^T + AY + \frac{1}{\gamma^2}YC_1^TC_1Y - (B_1D_{21}^T + YC_2^T)R_w^{-1}(D_{21}B_1^T + C_2Y) + B_1B_1^T = 0 \tag{8.122}$$

and L in (8.110) is given by

$$L = \begin{bmatrix} \frac{1}{\gamma^2}YC_1^T & -(B_1D_{21}^T + YC_2^T)R_w^{-1} \end{bmatrix} := \begin{bmatrix} L_z & L_y \end{bmatrix}. \tag{8.123}$$

Since

$$D_u^T(\gamma^{-2}D_wJ_\gamma^{-1}D_w)^{-1}D_u = \begin{bmatrix} R_z & 0 \\ 0 & -\gamma^{-2}R_w^{-1} \end{bmatrix},$$

V_w in (8.113) is written as

$$V_w = \begin{bmatrix} R_z^{1/2} & 0 \\ 0 & -\gamma^{-1}R_w^{-1/2} \end{bmatrix}. \tag{8.124}$$

The controller generators are given by

$$\Pi_{11} = V_w \left[\begin{array}{c|cc} A + L_1C_1 + L_2C_2 & -(B_2 + L_zD_{12}) & L_y \\ \hline \begin{bmatrix} F_u \\ C_2 + D_{21}F_w \end{bmatrix}\left(I - \frac{1}{\gamma^2}YX\right) & \begin{matrix} I \\ 0 \end{matrix} & \begin{matrix} 0 \\ I \end{matrix} \end{array}\right],$$

$$(8.125)$$

$$\Pi_{11}^{-1} = \left[\begin{array}{c|cc} A+B_1F_w+B_2F_u & (I-\gamma^{-2}YX)^{-1}\left[\, B_2+L_zD_{12} \right. & \left. -L_y \,\right] \\ \hline F_u & I & 0 \\ C_2+D_{21}F_w & 0 & I \end{array}\right] V_w^{-1}. \tag{8.126}$$

Now we summarize the result.

COROLLARY 8.14 *Assume that $D_{11} = 0$ in (8.1). Under the assumptions (A1) and (A2), the unnormalized H^∞ control problem is solvable iff the following conditions hold.*

(i) There exists a solution $X \geq 0$ of the Riccati equation (8.120) that stabilizes

$$\hat{A}_G = A + \frac{1}{\gamma^2}B_1B_1^TX - B_2R_z^{-1}(D_{12}^TC_1 + B_2^TX), \tag{8.127}$$

(ii) there exists a solution $Y \geq 0$ of the Riccati equation (8.122) that stabilizes

$$\hat{A}_H = A + \frac{1}{\gamma^2}YC_1^TC_1 - (B_1D_{21}^T + YC_2^T)R_w^{-1}C_2, \tag{8.128}$$

(iii) $\sigma(XY) < \gamma^2$.

In that case, the set of desirable controllers is given by (8.93) where Π_{11} and Π_{11}^{-1} are given by (8.125) and (8.126), respectively.

Remark : Condition (i) is equivalent to the existence of $X = Ric(H_z) \geq 0$, where

$$H_z := \begin{bmatrix} A & \frac{1}{\gamma^2}B_1B_1^T \\ -C_1^TC_1 & -A^T \end{bmatrix} + \begin{bmatrix} -B_2 \\ C_1^TD_{12} \end{bmatrix} R_z^{-1}\begin{bmatrix} D_{12}^TC_1 & B_2^T \end{bmatrix}. \tag{8.129}$$

Dually, Condition (ii) is equivalent to the existence of $Y = Ric(H_w) \geq 0$ where

$$H_w := \begin{bmatrix} A^T & \frac{1}{\gamma^2}C_1^TC_1 \\ -B_1B_1^T & -A \end{bmatrix} + \begin{bmatrix} -C_2^T \\ B_1D_{21}^T \end{bmatrix} R_w^{-1}\begin{bmatrix} D_{21}B_1^T & C_2 \end{bmatrix}. \tag{8.130}$$

(iv) The case where $D_{11} = 0, \quad D_{12}^T C_1 = 0, \quad B_1 D_{21}^T = 0.$
To see the meaning of these assumptions, consider a plant given by

$$\dot{x} = Ax + \bar{B}_1 w_1 + B_2 u, \tag{8.131a}$$
$$z_1 = \bar{C}_1 x, \tag{8.131b}$$
$$z_2 = \bar{D}_{12} u, \tag{8.131c}$$
$$y = C_2 x + \bar{D}_{21} w_2. \tag{8.131d}$$

The special features of this plant are twofold: the external signal w_1 driving the plant (8.131a) is separated from the external signal w_2 corrupting the measurement y in (8.131d). Also, the controlled variable z is divided into two parts as in (8.131b) and (8.131c) so that

$$\|z\|^2 = x^T C_1^T C_1 x + u^T \bar{D}_{12}^T \bar{D}_{12} u.$$

The plant (8.131) is described in the general form (8.1) with

$$C_1 = \begin{bmatrix} \bar{C}_1 \\ 0 \end{bmatrix}, \quad D_{12} = \begin{bmatrix} 0 \\ \bar{D}_{12} \end{bmatrix},$$
$$B_1 = \begin{bmatrix} \bar{B}_1 & 0 \end{bmatrix}, \quad D_{21} = \begin{bmatrix} 0 & \bar{D}_{21} \end{bmatrix}.$$

In this case, we have

$$D_{12}^T C_1 = 0, \quad B_1 D_{21}^T = 0. \tag{8.132}$$

Many practical plants have the form (8.131). The Riccati equation (8.120) and the gain (8.121) become in this case

$$XA + A^T X + \frac{1}{\gamma^2} X B_1 B_1^T X - X B_2 R_z^{-1} B_2^T X + C_1^T C_1 = 0, \tag{8.133}$$

$$F = \begin{bmatrix} \frac{1}{\gamma^2} B_1^T \\ -R_z^{-1} B_2^T \end{bmatrix} X = \begin{bmatrix} F_w \\ F_u \end{bmatrix}. \tag{8.134}$$

Similarly, the Riccati equation (8.122) and the gain (8.123) become

$$YA^T + AY + \frac{1}{\gamma^2} Y C_1^T C_1 Y - Y C_2^T R_w^{-1} C_2 Y + B_1 B_1^T = 0, \tag{8.135}$$

$$L = Y \begin{bmatrix} \frac{1}{\gamma^2} C_1^T & -C_2^T R_w^{-1} \end{bmatrix} = \begin{bmatrix} L_z & L_y \end{bmatrix}. \tag{8.136}$$

From (8.132), (8.134), and (8.136), it follows that $D_{21}F_w = 0$ and $L_z D_{12} = 0$. Thus, the controller generators are given by

$$\Pi_{11} = V_w \left[\begin{array}{c|cc} A+L_zC_1+L_yC_2 & -B_2 & L_y \\ \hline \begin{bmatrix} F_u \\ C_y \end{bmatrix}(I-\frac{1}{\gamma^2}YX) & \begin{matrix} I \\ 0 \end{matrix} & \begin{matrix} 0 \\ I \end{matrix} \end{array}\right], \tag{8.137}$$

$$\Pi_{11}^{-1} = \left[\begin{array}{c|cc} A+B_1F_w+B_2F_u & (I-\gamma^{-2}YX)^{-1}\begin{bmatrix} B_2 & -L_y \end{bmatrix} & \\ \hline F_u & I & 0 \\ C_y & 0 & I \end{array}\right] V_w^{-1}, \tag{8.138}$$

where V_w is given by (8.124). Now we summarize the reasoning.

COROLLARY 8.15 *Assume that $D_{11} = 0$ in (8.1) and the conditions (8.132) hold. Under the assumptions (A1) and (A2), the unnormalized H^∞ control problem is solvable iff the following conditions hold.*

(i) There exists a solution $X \geq 0$ of the Riccati equation (8.133) that stabilizes

$$\hat{A}_G = A + (\frac{1}{\gamma^2}B_1B_1^T - B_2R_z^{-1}B_2^T)X,$$

(ii) there exists a solution $Y \geq 0$ of the Riccati equation (8.135) that stabilizes

$$\hat{A}_H = A + Y(\frac{1}{\gamma^2}C_1^TC_1 - C_2^TR_w^{-1}C_2),$$

(iii) $\sigma(XY) < \gamma^2$.

In that case, the set of all desirable controllers is given by (8.93), where Π_{11} and Π_{11}^{-1} are given by (8.137) and (8.138), respectively.

Remark: This case was first treated in the well-known paper [24]. There, the plant was further simplified to $R_z = I$, $R_w = I$.

Notes

The state-space solution to the standard H^∞ control problem derived in this chapter is nothing but a straightforward application of the results of the preceding chapters except for the treatment of the augmentations.

It is interesting that the chain-scattering representations of the plant P are obtained either by a state feedback and an input transformation of a simpler plant (a partial plant P_z) or by an output insertion and an output transformation of another partial plant P_w. Implications of these structures need to be exploited further. The key result that deals with the augmentations is Theorem 8.8 which depends heavily on the scattering structure of the plant and Lemma 3.7. The maximum augmentation introduced there can be regarded as a class of input/output augmentations that do not contribute to enhance the performance of H^∞ control.

The solution to the standard H^∞ control problem was initially reported in [33] for a restricted class of plants without proof. The celebrated work [24] was the first paper that described the whole structure of H^∞ control systems in a comprehensible way to engineers. However, the class of plants treated in that paper was limited. The solution to the problem in the same generality as in Theorems 8.11 and 8.12 was initially given in a series of papers [34] based on the unitary dilation theory. A more compact derivation was given in [36] based on J-spectrum factorization which is similar to our approach. A different version of the chain-scattering representation was used in [96] to derive the solution to H^∞ control. The approach of [96] can bypass the augmentation using a factorization technique. Simplified derivations are found in [43][76][86][66][109][89][92].

The readers may notice that the assumption $D_{11} = 0$ simplifies the derivation significantly, as was noted in 8.5. In [85], it was shown that we can make this assumption without loss of generality. However, since the argument of [85] is itself heavy, we followed the most general line of reasoning.

Recently, the solution to nonstandard cases where the assumption (A2) does not hold was obtained in an elementary way based on the linear matrix inequality [30][44].

Problems

[1] Assume that both D_{12} and D_{21} are invertible. Show that C_G in (8.6) and B_H in (8.14) are given, respectively, by

$$C_G = -D_w^{-1}C, \quad B_H = BD_z^{-1}J.$$

[2] Under the same assumptions as Problem [1], show that the normalized H^∞ control problem is solvable iff the following conditions hold.

(i) The Riccati equation

$$\bar{X}A_G^T + A_G\bar{X} + \bar{X}C_G^TJC_G\bar{X} = 0$$

has a solution $\bar{X} \geq 0$ that stabilizes $A_G + \bar{X}C_G^TJC_G$.

(ii) The Riccati equation

$$XA_H + A_H^TX + XB_HJB_H^TX = 0$$

has a solution $X \geq 0$ that stabilizes $A_H + B_HJB_H^TX$.

(iii) $\sigma(\bar{X}X) < 1$,

where A_G and C_G are given in (8.6) with $\hat{D}_{21}$ and $\hat{C}_2$ being replaced by D_{21} and C_2, respectively, and A_H, B_H are given in (8.14) with $\hat{D}_{12}$ and $\hat{B}_2$ being replaced by D_{21} and B_2, respectively.

[3] Under the same assumption as Problem [1], prove that the following identities hold in Theorems 8.2 and 8.3.

$$X = \bar{Y}, \qquad Y = \bar{X},$$
$$\hat{A}_G = \bar{A}_H, \qquad \hat{A}_H = \bar{A}_G,$$
$$F_G = C_H - D_HJB_H^TX, \quad L_H = -B_G - C_G^TJD_G.$$

Using these identities, show that

$$\Pi(s)^{-1} = \Omega(s).$$

[4] Show that maximum augmentations are always admissible.

[5] Show that the inequality (8.48) holds iff

$$U := I - D_{11}^T(I - D_{12}(D_{12}^TD_{12})^{-1}D_{12}^T)D_{11} > 0.$$

Also, show that

$$E_z = \begin{bmatrix} (D_{12}^TD_{12})^{-1/2} & 0 \\ 0 & U^{1/2} \end{bmatrix} \begin{bmatrix} D_{12}^TD_{11} & D_{12}^TD_{12} \\ I & 0 \end{bmatrix}$$

satisfies (8.26).

[6] Show that the inequality (8.49) holds iff

$$V := I - D_{11}(I - D_{21}^T(D_{21}D_{21}^T)^{-1}D_{21})D_{11}^T > 0.$$

Also, show that

$$E_z = \begin{bmatrix} D_{11}^T D_{21} & I \\ D_{21}D_{21}^T & 0 \end{bmatrix} \begin{bmatrix} 0 & (D_{21}D_{21}^T)^{-1/2} \\ V^{1/2} & 0 \end{bmatrix}$$

satisfies (8.36).

[7] Show that the inequalities (8.105) and (8.106) are always satisfied for sufficiently large γ.

[8] Show that the Riccati equation (8.118) can be written as

$$\begin{aligned} & X(A+BF) + (A+BF)^T X \\ & \quad +(D_r F + C_1)^T(D_r F + C_1) - F^T \begin{bmatrix} \gamma^2 I & 0 \\ 0 & 0 \end{bmatrix} F = 0. \end{aligned} \tag{8.139}$$

[9] Let F in (8.106) be written as

$$F = \begin{bmatrix} F_w \\ F_u \end{bmatrix},$$

as in (8.119) and let $z = z_0(t)$ be the output of (8.1) corresponding to $w = w_0(t) = F_w x(t)$ and $u = u_0(t) = F_u x(t)$ with $x(0) = x_0$. Show that

$$\int_0^\infty (z_0^T(t)z_0(t) - \gamma^2 w_0^T(t)w_0(t))dt = -x_0^T X x_0,$$

where X is a solution to (8.139).

Chapter 9

Structure of H^∞ Control

9.1 Stability Properties

In this section, we prove the stability of some closed-loop matrices based on the Riccati equations (8.120) and (8.122), under the standing assumption

$$D_{11} = 0. \tag{9.1}$$

The Riccati equation (8.120) is written as

$$XA + A^TX + \frac{1}{\gamma^2}XB_1B_1^TX + XB_2F_u + C_1^T(C_1 + D_{12}F_u) = 0, \tag{9.2}$$

where F_u is given by (8.121). Using the identity $B_2^TX = -(R_zF_u + D_{12}^TC_1)$ and (8.119), we have an alternative representation of (9.2) as follows.

$$\begin{aligned} &X(A + B_2F_u) + (A + B_2F_u)^TX \\ &\quad + (C_1 + D_{12}F_u)^T(C_1 + D_{12}F_u) + \frac{1}{\gamma^2}XB_1B_1^TX = 0. \end{aligned} \tag{9.3}$$

If $X \geq 0$ is a stabilizing solution of (8.119), $A_G = A + B_1F_w + B_2F_u = A + \gamma^{-2}B_1B_1^TX + B_2F_u$ is stable. We show that $A + B_2F_u$ is also stable.

Let λ be an eigenvalue of $A + B_2F_u$ with x being the associated eigenvector; that is, $(A + B_2F_u)x = \lambda x$. From (9.3), it follows that

$$(\lambda + \bar{\lambda})x^*Xx + \|(C_1 + D_{12}F_u)x\|^2 + \frac{1}{\gamma^2}\|B_1^TXx\|^2 = 0.$$

Assume that $\lambda + \bar{\lambda} \geq 0$. Then we have $Xx = 0$. Hence, $\hat{A}_Gx = (A + \gamma^{-2}B_1B_1^TX + B_2F_u)x = \lambda x$. This contradicts the stability of $\hat{A}_G$.

Therefore, $\lambda + \bar{\lambda} < 0$, which implies that $A + B_2 F_u$ is stable. Thus, we have established the following result.

LEMMA 9.1 *If the Riccati equation (8.120) has a solution $X \geq 0$ which stabilizes $\hat{A}_G = A + B_1 F_w + B_2 F_u$, that is, $X = Ric(H_z) \geq 0$ exists, then $A + B_2 F_u$ is also stable.*

The dual of (9.3) is derived from (8.122) as

$$Y(A + L_y C_2)^T + (A + L_y C_2)Y + (B_1 + L_y D_{21})(B_1 + L_y D_{21})^T + \frac{1}{\gamma^2} Y C_1^T C_1 Y = 0. \qquad (9.4)$$

Based on this relation, we can establish the dual of Lemma 9.1.

LEMMA 9.2 *If the Riccati equation (8.122) has a solution $Y \geq 0$ which stabilizes $\hat{A}_H = A + L_z C_1 + L_y C_2$, that is, $X = Ric(H_w) \geq 0$ exists, then $A + L_y C_2$ is stable.*

Now, we derive another stable matrix for later use.

LEMMA 9.3 *Assume that the conditions (i)~(iii) of Corollary 8.14 hold. Then $Z := (I - \gamma^{-2} Y X)^{-1} Y \geq 0$ is a solution of the Riccati equation*

$$Z(A + B_1 F_w)^T + (A + B_1 F_w)Z + \frac{1}{\gamma^2} Z F_u^T R_z F_u Z + B_1 B_1^T - U L_y R_w L_y^T U^T = 0, \qquad (9.5)$$

such that

$$\hat{A}_1 := A + B_1 F_w + \frac{1}{\gamma^2} Z F_u^T R_z F_u + U L_y (C_2 + D_{21} F_w) \qquad (9.6)$$

is stable, where

$$U = (I - \gamma^{-2} Y X)^{-1}. \qquad (9.7)$$

Proof. Let $X = \text{Ric}\ (H_z)$ and let

$$T := \begin{bmatrix} I & -\gamma^{-2} X \\ 0 & I \end{bmatrix}.$$

Direct computation using (8.130) and (8.121) yields

$$TH_wT^{-1} = \begin{bmatrix} (A+B_1F_w)^T & \gamma^{-2}F_u^TR_zF_u \\ -B_1B_1^T & -(A+B_1F_w) \end{bmatrix} - \begin{bmatrix} C_2^T + F_w^TD_{21}^T \\ -B_1D_{21}^T \end{bmatrix} R_w^{-1} \begin{bmatrix} D_{21}B_1^T & C_2 + D_{21}F_w \end{bmatrix}.$$

which is obviously Hamiltonian (See Problem 3.1). Due to a well-known property of the Riccati equation, $Z = \text{Ric}\,(TH_wT^{-1})$ (See Problem 3.1). Hence,

$$\begin{bmatrix} Z & -I \end{bmatrix} TH_wT^{-1} \begin{bmatrix} I \\ Z \end{bmatrix} = 0.$$

From $D_{21}B_1^T(I-\gamma^{-2}XY)+(C_2+D_{21}F_w)Y = D_{21}B_1^T + C_2Y = -R_wL_y^T$, we have the identity $D_{21}B_1^T + (C_2+D_{21}F_w)Z = -R_wL_y^TU^T$. Using these relations, we can show (9.5) immediately. ∎

Based on the preceding lemma, we can show the following result.

LEMMA 9.4 *Assume that the conditions (i)~(iii) of Corollary 8.14 hold. Then*

$$\hat{A} := A + B_1F_w + UL_y(C_2 + D_{21}F_w) \tag{9.8}$$

is stable.

Proof. From (8.123), and the relation $(C_2 + D_{21}F_w)Z = -(D_{21}B_1^T + R_wL_y^TU^T)$, it follows that $UL_yR_wL_y^TU^T = -UL_y(C_2+D_{21}F_w)Z - Z(C_2+D_{21}F_w)^TL_y^TU^T + B_1D_{21}R_w^{-1}D_{21}^TB_1^T - Z(C_2 + D_{21}F_w)^T(C_2 + D_{21}F_w)Z$. Substitution of this relation into (9.5) yields

$$Z\hat{A}^T + \hat{A}Z + B_1(I - D_{21}^TR_w^{-1}D_{21})B_1^T + \frac{1}{\gamma^2}ZF_u^TR_zF_uZ + Z(C_2 + D_{21}F_w)^T(C_2 + D_{21}F_w)Z = 0.$$

Since $\hat{A}_1$ given by (9.6) is stable, $(\hat{A}, R_zF_u)$ is detectable. Since $I - D_{21}R_w^{-1}D_{21}^T \geq 0$, the assertion follows from the fact shown in Problem 3.6. ∎

9.2 Closed-Loop Structure of H^∞Control

This section is devoted to the investigation of the closed-loop structure of an H^∞ control system. We focus on the case $D_{11} = 0$ (the third case in Section 8.5), and show that the controller is actually of the form that combines state feedback control law with the state estimator, as in the *LQG* case.

The plant is now given by

$$\dot{x} = Ax + B_1 w + B_2 u, \tag{9.9a}$$
$$z = C_1 x + D_{12} u, \tag{9.9b}$$
$$y = C_2 x + D_{21} w. \tag{9.9c}$$

Let $X \geq 0$ be a solution to the algebraic Riccati equation (8.120) and write

$$\psi(x) = x^T X x. \tag{9.10}$$

The differential of $\psi(x(t))$ along the trajectory of (9.9a) is calculated to be

$$\begin{aligned}\frac{d}{dt}\psi(x(t)) &= \dot{x}^T X x + x^T X \dot{x}\\ &= (Ax + B_1 w + B_2 u)^T X x + x^T X(Ax + B_1 w + B_2 u)\\ &= x^T(A^T X + XA)x + \gamma^2(x^T F_w^T w + w^T F_w x)\\ &\qquad + u^T B_2^T X x + x^T X B_2 u.\end{aligned}$$

Using (8.120) and (8.121), we obtain

$$\begin{aligned}\frac{d}{dt}\psi(x(t)) &= -x^T(\gamma^2 F_w^T F_w - F_u^T R_z F_u + C_1^T C_1)x\\ &\qquad + \gamma^2(x^T F_w^T w + w^T F_w x) - u^T(R_z F_u + D_{12}^T C_1)x\\ &\qquad - x^T(F_u^T R_z + C_1^T D_{12})u\\ &= -\gamma^2(w - F_w x)^T(w - F_w x) + \gamma^2 w^T w\\ &\qquad + (u - F_u x)^T R_z(u - F_u x)\\ &\qquad - (C_1 x + D_{12}u)^T(C_1 x + D_{12}u)\\ &= -z^T z + \gamma^2 w^T w - \gamma^2(w - F_w x)^T(w - F_w x)\\ &\qquad + (u - F_u x)^T R_z(u - F_u x).\end{aligned}$$

Integration of both sides in the interval $[\ 0 \quad T\ ]$ yields

$$x^T(T)Xx(T) - x^T(0)Xx(0) = \int_0^T (\gamma^2\|w(t)\|^2 - \|z(t)\|^2)dt$$

$$-\gamma^2 \int_0^T \|w(t) - F_w x(t)\|^2 dt + \int_0^T \|R_z^{1/2}(u(t) - F_u x(t))\|^2 dt.$$

Taking $x(0) = 0$ gives

$$\begin{aligned} J(w,u;T) &:= \int_0^T (\gamma^2\|w(t)\|^2 - \|z(t)\|^2)dt \\ &= x^T(T)Xx(T) + \gamma^2 \int_0^T \|w(t) - F_w x(t)\|^2 dt \\ &\quad - \int_0^T \|R_z^{1/2}(u(t) - F_u x(t))\|^2 dt. \end{aligned} \tag{9.11}$$

This is a key identity from which various properties of H^∞control can be extracted. Obviously, $J(w,u;T) \to \gamma^2\|w\|^2 - \|z\|^2$, as $T \to \infty$. The design objective (7.5) is attained iff

$$J(w,u;\infty) > 0, \ \forall w \in L_2. \tag{9.12}$$

Actually, if we can implement the state feedback control law

$$u_0 = F_u x, \tag{9.13}$$

the design objective (9.12) is attained with possible equality due to $X \geq 0$.

These results can be verified alternatively. The state feedback control law (9.13) applied to the plant (9.9) yields the closed-loop transfer function from w to z as

$$T_{zw}(s) = (C_1 + D_{12}F_u)(sI - A - B_2F_u)^{-1}B_1.$$

According to Corollary 3.11, the existence of the solution $X \geq 0$ of (9.3) implies

$$\|T_{zw}(s)\|_\infty \leq \gamma.$$

Thus the design objective is achieved by the state feedback (9.13).

LEMMA 9.5 *The state feedback (9.13) achieves the design objective for the plant (9.9).*

On the other hand, the exogenous signal given by

$$w_0 = F_w \xi \tag{9.14}$$

represents the worst case, in the sense that it maximizes $J(w, u; \infty)$. Now we investigate the closed-loop structure of H^∞control. As was shown in Theorem 8.12, a desirable H^∞controller is given by

$$K = HM(\Pi_{11}^{-1}; S), \tag{9.15}$$

where S is an arbitrary matrix in $\mathbf{BH}^\infty$ and

$$\Pi_{11}^{-1} = \left[\begin{array}{c|cc} A + B_1F_w + B_2F_u & U\left[\begin{array}{cc} B_2 + L_zD_{12} & -L_y \end{array}\right] \\ \hline F_u & I & 0 \\ C_2 + D_{21}F_w & 0 & I \end{array}\right] V_w^{-1}. \tag{9.16}$$

Since V_w is given by (8.124), Π_{11}^{-1} is described in the state space as

$$\begin{aligned} \dot{\xi} &= (A + B_1F_w + B_2F_u)\xi + \gamma U(B_2 + L_zD_{12})R_z^{-1/2}a \\ &\quad + UL_yR_w^{1/2}b, && (9.17a) \\ u &= F_u\xi + \gamma R_z^{-1/2}a, && (9.17b) \\ y &= (C_2 + D_{21}F_w)\xi - R_w^{1/2}b, && (9.17c) \end{aligned}$$

where ξ is the state of the controller and U is given by (9.7). The controller (9.14) is obtained by introducing the relation

$$a = S(s)b.$$

The controller can be rewritten as

$$\begin{aligned} \dot{\xi} &= (A+B_1F_w+B_2F_u)\xi \\ &\quad +U\left[(B_2+L_zD_{12})\bar{S}(s)+L_y\right]\nu, && (9.18a) \\ u &= F_u\xi + \bar{S}(s)\nu, && (9.18b) \\ \nu &= y - (C_2 + D_{21}F_w)\xi, && (9.18c) \end{aligned}$$

where $\bar{S}; = \gamma R_z^{-1/2}S(s)R_w^{-1/2}$. The block-diagram of the above controller is illustrated in Figure 9.1. If we choose $\bar{S}(s) = 0$ in (9.18), we obtain the so-called *central controller*, which is described as

$$\begin{aligned} \dot{\xi} &= A\xi + B_1\hat{w}_0 + B_2u - UL_y(y - C_2\xi - D_{21}\hat{w}_0), && (9.19a) \\ u &= F_u\xi, && (9.19b) \end{aligned}$$

where

$$\hat{w}_0 = F_w\xi. \tag{9.20}$$

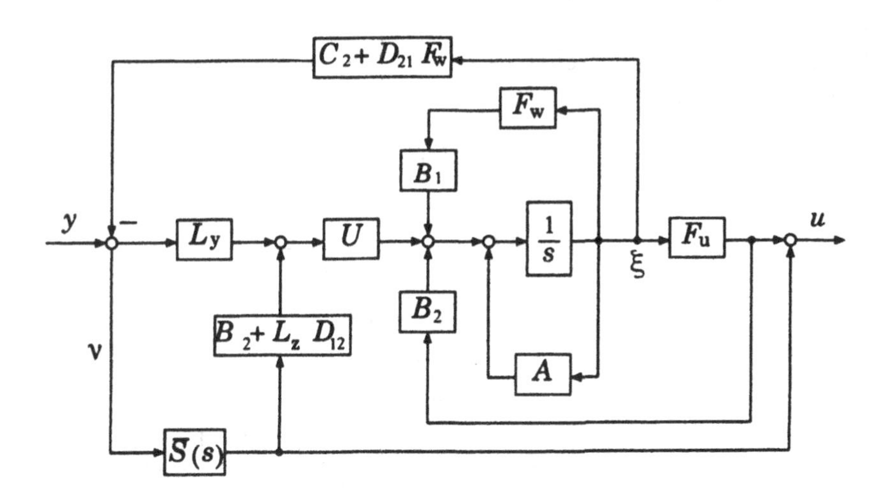

Figure 9.1 Block Diagram of H^∞ Controller.

The representation (9.19a) clarifies the observer structure of the central controller. The control law (9.19b) is just the replacement of x by ξ in (9.13). Hence, (9.20) represents the estimate of the state feedback control law. The most interesting feature of (9.19a) is that it assumes the exogenous signal w to be the worst. The signal $\hat{w}_0$ given by (9.20) represents the estimate of the worst one. In view of (9.9c), $\nu = y - C_2\xi - D_{21}\hat{w}_0$ represents the innovation assuming $w = \hat{w}_0$. The observer gain is given by UL_y. Figure 9.2 illustrates the block diagram of the central controller.

The central controller is regarded as an observer-based *quasi-state feedback* assuming that the exogenous signal is worst, that is, assuming the plant dynamics

$$\dot{x} = (A + B_1F_w)x + B_2u, \tag{9.21a}$$
$$y = (C_2 + D_{12}F_w)x. \tag{9.21b}$$

Substituting (9.20) in (9.19a) yields

$$\dot{\xi} = (A + B_1F_w)\xi + B_2u - UL_y(y - (C_2 + D_{21}F_w)\xi). \tag{9.22}$$

which is a usual identity observer for the plant (9.21) with observer gain UL_y. Lemma 9.4 shows that the gain UL_y stabilizes $(A + B_1F_w) + UL_y(C_2 + D_{21}F_w)$, which is certainly necessary for UL_y to be an observer gain. The control input (9.19b) is clearly the estimate of the control law (9.13) via the observer (9.19). In the case where the condition (8.132) holds, the central controller (9.21), (9.19a) is further simplified due to the obvious relations

$$D_{21}F_w = 0, \quad F_u = -R_z^{-1}B_2^T X, \quad L_y = -YC_2^T R_w^{-1}.$$

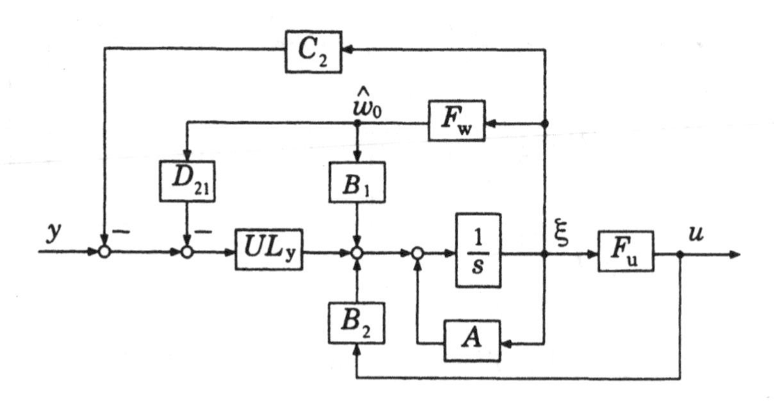

Figure 9.2 Block Diagram of Central Controller.

9.3 Examples

In this section, a collection of simple examples is given, which can be calculated by hand, in order to get an idea of the structure of H^∞control.

Example 9.1
Consider a first-order plant

$$\begin{aligned}
\dot{x} &= ax + b_1 w + b_2 u, && (9.23a)\\
z &= c_1 x + d_{12} u, && (9.23b)\\
y &= c_2 x + d_{21} w, && (9.23c)
\end{aligned}$$

where all the quantities in these expressions are scalar. The assumption (A1) implies that

$$a < 0 \quad or \quad b_2 c_2 \neq 0.$$

The assumption (A2) implies that

$$d_{12}d_{21} \neq 0, \quad (a - \frac{b_2 c_1}{d_{12}})(a - \frac{b_1 c_2}{d_{21}}) \neq 0.$$

The Riccati equation (8.107) (or (8.120)) becomes in this case

$$2\left(a-\frac{b_2c_1}{d_{12}}\right)X-\left(\frac{b_2^2}{d_{12}^2}-\frac{b_1^2}{\gamma^2}\right)X^2=0.$$

The stabilizing solution $X \geq 0$ exists iff $\beta_c > 0$, and is given by

$$X=\begin{cases} 0, & \text{if} \quad \alpha_c < 0, \\ \dfrac{2\alpha_c}{\beta_c}, & \text{if} \quad \alpha_c > 0, \end{cases}$$

where

$$\alpha_c := a-\frac{b_2c_1}{d_{12}}, \quad \beta_c := \frac{b_2^2}{d_{12}^2}-\frac{b_1^2}{\gamma^2},$$

and the stabilized matrix is given by

$$A+BF=\begin{cases} \alpha_c, & \text{if} \quad \alpha_c < 0, \\ -\alpha_c, & \text{if} \quad \alpha_c > 0. \end{cases}$$

It is important to notice that the role of F is to bring the closed-loop pole at the mirror image of $a - b_2c_1/d_{12}$, the zero of $P_{12}(s)$. Dually, the Riccati equation (8.109)(or (8.138)) becomes in this case

$$2\left(a-\frac{b_1c_2}{d_{21}}\right)Y-\left(\frac{c_2^2}{d_{21}^2}-\frac{c_1^2}{\gamma^2}\right)Y^2=0.$$

The stabilizing solution Y exists iff $\beta_o > 0$, and is given by

$$Y=\begin{cases} 0, & \text{if } \alpha_o < 0, \\ \dfrac{2\alpha_o}{\beta_o}, & \text{if } \alpha_o > 0, \end{cases}$$

where

$$\alpha_o = a-\frac{b_2c_1}{d_{21}}, \quad \beta_o = \frac{c_2^2}{d_{21}^2}-\frac{c_1^2}{\gamma^2},$$

and the stabilized matrix (8.110) is given by

$$A+LC=\begin{cases} \alpha_o, & \text{if} \quad \alpha_o < 0, \\ -\alpha_o, & \text{if} \quad \alpha_o > 0. \end{cases}$$

The solvability condition holds iff one of the following conditions holds.

(i) $\alpha_c < 0$, $\alpha_o < 0$,

(ii) $\alpha_c < 0$, $\alpha_o > 0$, $\beta_o > 0$,

(iii) $\alpha_c > 0$, $\alpha_o < 0$, $\beta_c > 0$, or

(iv) $\alpha_c > 0$, $\alpha_o > 0$, $\beta_c > 0$, $\beta_o > 0$, $\gamma^2\beta_o\beta_c > 4\alpha_o\alpha_c$.

The central controller is given by

$$\dot{\xi} = a\xi + b_1\hat{w} + b_2u + \rho_\infty(y - c_2\xi - d_{21}\hat{w}), \tag{9.24a}$$

$$u = f_\infty\xi, \tag{9.24b}$$

$$\hat{w} = k_\infty\xi, \tag{9.24c}$$

where the three gains f_∞, k_∞, and ρ_∞ are given in Table 9.1.

Case	f_∞	k_∞	ρ_∞
(i)	$-\frac{c_1}{d_{12}}$	0	$\frac{b_1}{d_{21}}$
(ii)	$-\frac{c_1}{d_{12}}$	0	$\frac{b_1}{d_{21}} + \frac{2c_2\alpha_o}{d_{21}^2\beta_o}$
(iii)	$-(\frac{c_1}{d_{12}} + \frac{2b_2\alpha_c}{\beta_c})$	$\frac{2b_1\alpha_c}{\gamma^2\beta_c}$	$\frac{b_1}{d_{21}}$
(iv)	$-(\frac{c_1}{d_{12}} + \frac{2b_2\alpha_c}{\beta_c})$	$\frac{2b_1\alpha_c}{\gamma^2\beta_c}$	$\frac{\gamma^2\beta_c\beta_o(b_1d_{21} + 2c_2\alpha_o)}{(\gamma^2\beta_c\beta_o - 4\alpha_c\alpha_o)d_{21}^2\beta_o}$

Table 9.1 List of Gains of the Controller (9.24)

Example 9.2
Consider another first-order system given by

$$\dot{x} = ax + w_1 + b_2u, \tag{9.25a}$$

$$z_1 = x, \tag{9.25b}$$

$$z_2 = \sigma u, \tag{9.25c}$$

$$y = c_2x + d_{21}w_2. \tag{9.25d}$$

This is an example that satisfies the condition (8.132). The design objective is to satisfy

$$\|z\|^2 = \int_0^\infty (x(t)^2 + \sigma^2u(t)^2)dt \le \gamma^2\|w\|^2 = \gamma^2\int_0^\infty w(t)^2dt.$$

The Riccati equation (8.133) becomes in this case

$$2aX + \left(\frac{1}{\gamma^2} - \frac{b_2^2}{\sigma^2}\right) X^2 + 1 = 0. \tag{9.26}$$

This equation has a solution $X \geq 0$ such that $a + \delta X < 0$ iff

$$a^2 > \delta \quad \text{and} \quad a\delta < 0, \tag{9.27}$$

where

$$\delta := \frac{1}{\gamma^2} - \frac{b_2^2}{\sigma^2}.$$

The Riccati equation (8.135) becomes

$$2aY + \left(\frac{1}{\gamma^2} - \frac{c_2^2}{d_{21}^2}\right) Y^2 + 1 = 0. \tag{9.28}$$

This equation has a solution $Y \geq 0$ such that $a + \varepsilon Y < 0$, iff

$$a^2 > \varepsilon, \quad a\varepsilon < 0, \tag{9.29}$$

where

$$\varepsilon := \frac{1}{\gamma^2} - \frac{c_2^2}{d_{21}^2}.$$

The inequality (8.111) becomes in this case

$$\left(a + \sqrt{a^2 - \delta}\right)\left(a + \sqrt{a^2 - \varepsilon}\right) < \gamma^2 \delta \varepsilon. \tag{9.30}$$

The solvability condition is given by (9.27), (9.29), and (9.30). The central controller is given by

$$\dot{\xi} = a\xi + \hat{w}_1 + b_2 u + \rho_\infty (y - c_2 \xi - d_{21}\hat{w}), \tag{9.31a}$$

$$u = f_\infty \xi, \tag{9.31b}$$

$$\hat{w}_1 = k_\infty \xi, \tag{9.31c}$$

$$\begin{aligned} f_\infty &= -\frac{ab_2}{\sigma^2 \delta}\left(1 + \sqrt{1 - \frac{\delta}{a^2}}\right), \\ k_\infty &= \frac{ab_1}{\gamma^2 \delta}\left(1 + \sqrt{1 - \frac{\delta}{a^2}}\right), \\ \rho_\infty &= \frac{c_2 \gamma^2 \delta}{d_{21}^2} \frac{a + \sqrt{a^2 - \varepsilon}}{\gamma^2 \delta \varepsilon - \left(a + \sqrt{a^2 - \delta}\right)\left(a + \sqrt{a^2 - \varepsilon}\right)}. \end{aligned} \tag{9.31d}$$

For comparison, the LQG controller is given by

$$\begin{aligned}\dot{\xi} &= a\xi + b_2 u + \rho_2(y - c_2\xi),\\ u &= f_2\xi,\end{aligned}$$

where

$$f_2 = -\frac{1}{b_2}\left(a + \sqrt{\frac{a^2}{b_2^2} + \frac{1}{\sigma^2}}\right),$$

$$\rho_2 = \frac{1}{c_2}\left(a + \sqrt{a^2 + \frac{c_2^2}{d_{21}^2}}\right).$$

Example 9.3
Finally, we consider a second-order system

$$\dot{x}_1 = x_2, \tag{9.32a}$$
$$\dot{x}_2 = w_1 + u, \tag{9.32b}$$
$$z_1 = x_1, \tag{9.32c}$$
$$z_2 = \sigma u, \tag{9.32d}$$
$$y = c_2 x_1 + d_{21} w_2, \quad \sigma > 0. \tag{9.32e}$$

The assumptions (A1)~(A3) are satisfied in this case. The Riccati equation (8.107) becomes

$$X\begin{bmatrix} 0 & 1 \\ 0 & 0 \end{bmatrix} + \begin{bmatrix} 0 & 0 \\ 1 & 0 \end{bmatrix} X + X\begin{bmatrix} 0 & 0 \\ 0 & \gamma^{-2}-\sigma^{-2} \end{bmatrix} X + \begin{bmatrix} 1 & 0 \\ 0 & 0 \end{bmatrix} = 0.$$

It is not difficult to see that this Riccati equation has a stabilizing solution $X \geq 0$, iff

$$\gamma > \sigma. \tag{9.33}$$

Under this condition, the stabilizing solution is given by

$$X = \begin{bmatrix} \sqrt{2\delta} & \delta \\ \delta & \sqrt{2\delta}\cdot\delta \end{bmatrix}, \quad \frac{1}{\delta^2} = \frac{1}{\sigma^2} - \frac{1}{\gamma^2}.$$

Dually, the Riccati equation (8.109) becomes

$$Y\begin{bmatrix} 0 & 0 \\ 1 & 0 \end{bmatrix} + \begin{bmatrix} 0 & 1 \\ 0 & 0 \end{bmatrix} Y + Y\begin{bmatrix} \gamma^{-2}-(c_2/d_{21})^2 & 0 \\ 0 & 0 \end{bmatrix} Y + \begin{bmatrix} 0 & 0 \\ 0 & 1 \end{bmatrix} = 0.$$

This Riccati equation has a stabilizing solution $Y \geq 0$ iff

$$\gamma > |d_{21}/c_2|.$$

Under this condition, the stabilizing solution is given by

$$Y = \begin{bmatrix} \sqrt{2\varepsilon}\cdot\varepsilon & \varepsilon \\ \varepsilon & \sqrt{2\varepsilon} \end{bmatrix}, \quad \frac{1}{\varepsilon^2} = \frac{c_2^2}{d_{21}^2} - \frac{1}{\gamma^2}.$$

The inequality (8.111) is represented as

$$\sqrt{\delta\varepsilon(\delta+\varepsilon)}\left(\sqrt{\delta}+\sqrt{\varepsilon}+\sqrt{\delta+\varepsilon}\right)+\delta\varepsilon < \gamma^2.$$

The central controller is given by

$$\begin{aligned}
\dot{\xi} &= \begin{bmatrix} 0 & 1 \\ 0 & 0 \end{bmatrix}\xi + \begin{bmatrix} 0 \\ 1 \end{bmatrix}(\hat{w}_1 + u) + \rho_\infty(y - c_2\xi - d_{21}\hat{w}_2), \\
u &= f_\infty \xi, \\
\hat{w}_1 &= k_{1\infty}\xi, \\
f_\infty &= -\frac{\delta}{\sigma^2}\begin{bmatrix} 1 & \sqrt{2\delta} \end{bmatrix}, \\
k_{1\infty} &= \frac{\delta}{\gamma^2}\begin{bmatrix} 1 & \sqrt{2\delta} \end{bmatrix}, \\
\rho_\infty &= (I - \gamma^{-2}YX)^{-1}\begin{bmatrix} \sqrt{2\varepsilon} \\ 1 \end{bmatrix}\frac{c_2\varepsilon}{d_{21}^2}.
\end{aligned}$$

For comparison, the *LQG* controller is given by

$$\begin{aligned}
\dot{\xi} &= \begin{bmatrix} 0 & 1 \\ 0 & 0 \end{bmatrix}\xi + \begin{bmatrix} 0 \\ 1 \end{bmatrix}u + \rho_2(y - c_2\xi), \\
u &= f_2\xi, \\
f_2 &= -\begin{bmatrix} \frac{1}{\sigma} & \sqrt{\frac{2}{\sigma}} \end{bmatrix}, \\
\rho_2 &= \begin{bmatrix} \sqrt{2/d_{21}} \\ 1/d_{21} \end{bmatrix}.
\end{aligned}$$

Notes

The closed-loop structure of H^∞ control systems has been fully exploited in [24] under the condition (8.132). An earlier result that recognized the role of estimators in H^∞ control systems is found in [48]. The state feedback control law (9.13) was first derived in [49]. An interesting observation concerning the state feedback H^∞ control is found in [87]. The identity (9.11) which represents the close tie between H^∞ control and differential games was found by various authors [70][9][37][93].

Here we note some extensions of the approach taken in this book. Extensions of the chain-scattering representations to nonstandard cases were treated in [102][103] based on the descriptor form. A discrete-time version has now been done in [64]. Some results on nonlinear chain-scattering representations have been published in [7][8]. Also, the time-varying version is now being investigated actively by Lee et al. [68][67].

Problems

[1] Show that the Riccati equation (9.3) can be written as

$$\begin{aligned} &X(A+BF)+(A+BF)^TX \\ &\quad +(C_1+D_{12}F_u)^T(C_1+D_{12}F_u)-\gamma^2F_w^TF_w=0. \end{aligned} \tag{9.34}$$

[2] Let

$$U(s)=\left[\begin{array}{c|c} A+BF & B \\ \hline C_z+D_zF & D_z \end{array}\right].$$

Using (9.34), show that

$$U(s)^\sim\begin{bmatrix} I & 0 \\ 0 & -\gamma^2I \end{bmatrix}U(s)=\begin{bmatrix} I & 0 \\ 0 & -\gamma^2I \end{bmatrix}. \tag{9.35}$$

[3] Using the result of Problem [2], prove that

$$\int_0^\infty(\|z\|^2-\gamma^2\|w\|^2)dt=\int_0^\infty(\|u-u_0\|^2-\gamma^2\|w-w_0\|^2)dt,$$

where $u_0=F_ux$ and $w_0=F_wx$.

[4] Show that the identity (9.3) implies that

$$\|(C_1 + D_{12}F_u)(sI - A - B_2F_u)^{-1}B_1\|_\infty < \gamma.$$

Consider the implication of this inequality.

[5] Show that, if an input augmentation (B_2', D_{12}') is maximum, that is, if it satisfies (8.82), the following identity holds:

$$\begin{aligned} x^T(T)Xx(T) - x^T(0)Xx(0) = & \int_0^T (\gamma^2\|w(t)\|^2 - \|z(t)\|^2)dt \\ & -\gamma^2 \int_0^T (\|w(t)\|^2 - F_w\|x(t)\|^2)dt \\ & + \int_0^T \|(\hat{D}_{12}^T\hat{D}_{12})^{1/2}(\hat{u}(t) - \hat{F}_u x(t))\|^2 dt, \end{aligned}$$

where $\hat{D}_{12} = \begin{bmatrix} D_{12}' & D_{12} \end{bmatrix}$, $\hat{u}(t) = \begin{bmatrix} u'(t)^T & u(t)^T \end{bmatrix}^T$ and $\hat{F}_u = \begin{bmatrix} 0 & F_u^T \end{bmatrix}^T$.

REFERENCES

[1] V. M. Adamjan, D. Z. Arov and M. G. Krein. Infinite block Hankel matrices and related extension problems. *American Mathematical Society Translations*, 111:133–156, 1978.

[2] L. Ahlfors. *Complex Analysis.* McGraw-Hill, New York, 1996.

[3] J. A. Ball and N. Cohen. Sensitivity minimization in an H^∞ norm: Parameterization of all sub-optimal solutions. *International Journal of Control*, 46:785–816, 1987.

[4] J. A. Ball, I. Gohberg and L. Rodman. *Interpolation of Rational Matrix Functions*, Vol. 45 of *Operator Theory: Advances and Appl.* Birkhäuser Verlag, Boston, 1990.

[5] J. A. Ball and J. W. Helton. Lie group over the field of rational functions, signed spectral factorization, signed interpolation and amplifier design. *Journal of Operator Theory*, 8:19–64, 1982.

[6] J. A. Ball and J. W. Helton. A Beuling-Lax theorem for the Lie group $U(m,n)$ which contains most classical interpolation theory. *Journal of Operator Theory*, 9:107–142, 1983.

[7] L. Baramov and H. Kimura. Nonlinear L_2-gain suboptimal control. In *Proceedings of the 3rd IFAC Symposium on Nonlinear Control Systems Design, Tahoe City*, pages 221–226, 1995.

[8] L. Baramov and H. Kimura. Nonlinear J-lossless conjugation and factorization. to appear in *International Journal of Control*, 1996.

[9] T. Basar and P. Bernhard. *H^∞ Optimal Control and Related Minimax Design Problems: A Dynamic Game Approach, Systems and Control.* Birkhäuser, Boston, 1991.

[10] S. Bittanti (Ed.). *The Riccati Equation in Control, Systems, and Signals.* Lecture Notes, Pitagora Editrice, Bologna, 1989.

[11] R. W. Brockett. *Finite Dimensional Linear Systems.* John Wiley and Sons, New York, 1970.

[12] B. C. Chang and J. B. Pearson. Optimal disturbance reduction in linear multivariable systems. *IEEE Trans. Automatic Control*, 29:880–887, 1984.

[13] S. Darlington. Synthesis of reactance 4 poles which produce prescribed insertion loss characteristics including special applications to filter design. *Journal of Math. and Phys.*, 18:257–353, 1939.

[14] C. Davis, W. M. Kahan and H. F. Weinberger. Norm-preserving dilations and their applications to optimal error bounds. *SIAM Journal of Control and Optimization*, 20:445–469, 1982.

[15] P. H. Delsarte, Y. Genin and Y. Kamp. The Nevanlinna-Pick problem for matrix-valued functions. *SIAM Journal on Applied Math.*, 36:47–61, 1979.

[16] P. H. Delsarte, Y. Genin and Y. Kamp. On the role of the Nevanlinna-Pick problem in circuit theory. *International Journal of Circuit Theory and Appl.*, 9:177–187, 1981.

[17] P. Dewilde and H. Dym. Lossless chain-scattering matrices and optimum linear prediction; the vector case. *International Journal of Circuit Theory and Appl.*, 9:135–175, 1981.

[18] P. Dewilde and H. Dym. Lossless inverse scattering, digital filters and estimation theory. *IEEE Trans. on Information Theory*, 30:644–622, 1984.

[19] P. Dewilde, A. Vieira and T. Kailath. On a generalized Szegö-Levinson realization algorithm for optimal linear predictors based on a network synthesis approach. *IEEE Trans. on Circuit and Systems*, 25:663–675, 1978.

[20] P. Dorato and Y. Li. A modification of the classical Nevanlinna-Pick interpolation algorithm with applications to robust stabilization. *IEEE Trans. Automatic Control*, 31:645–648, 1986.

[21] J. C. Doyle. Analysis of feedback systems with structured uncertainties. In *Proceedings of the IEEE, Part D*, 129:242-250, 1982.

[22] J. C. Doyle. *Lecture Notes in Advanced Multivariable Control.* Lecture Note. ONR/Honeywell Workshop, Minneapolis, 1984.

[23] J. C. Doyle, B. A. Francis and A. R. Tannenbaum. *Feedback Control Theory.* Macmillan Publishing Company, 1992.

[24] J. C. Doyle, K. Glover, P. P. Khargonekar and B. A. Francis. State-space solutions to standard H_2 and H_∞ control problems. *IEEE Trans. Automatic Control*, 34(8):831–847, 1989.

[25] J. C. Doyle and G. Stein. Multivariable feedback design; concepts for a classical/modern synthesis. *IEEE Trans. Automatic Control*, 26:4–16, 1981.

[26] H. Dym. *J-Contractive Matrix Functions, Reproducing Kernel Hilbert Spaces and Interpolation*, Vol. 71 of *Regional Conference Series in Math.* American Math. Soc. Providence, R.I., 1989.

[27] I. P. Fedcina. A criterion for the solvability of the Nevanlinna-Pick tangencial problem. *Mat. Isshedraniya*, Vol. 7:213–227, 1972.

[28] B. A. Francis. *A Course in H^∞ Control.* Lecture Notes in Control and Information Science. Springer Verlag, Toronto, 1987.

[29] B. A. Francis and G. Zames. On H^∞-optimal sensitivity theory for SISO feedback systems. *IEEE Trans. Automatic Control*, 29:9–16, 1984.

[30] P. Gahinet and P. Apkarian. A linear matrix in equality approach to H^∞ control. *International Journal of Robust and Nonlinear Control*, 4:421–448, 1994.

[31] K. Glover. All optimal Hankel-norm approximations of linear multivariable systems and their L^∞-error bounds. *International Journal of Control*, 39:1115–1193, 1984.

[32] K. Glover. Robust stabilization of linear multivariable systems: Relations to approximation. *International Journal of Control*, 43:741–766, 1986.

[33] K. Glover and J. C. Doyle. State space formulae for all stabilizing controllers that satisfy a H_∞ norm bound and relations to risk sensitivity. *Systems and Control Letters*, 11:167–172, 1988.

[34] K. Glover, D. J. N. Limebeer, J. C. Doyle, E. M. Kasenally and M. G. Safonov. A characterization of all solutions to the four block general distance problem. *SIAM Journal of Control and Optimization*, 29:283–324, 1991.

[35] M. Green. H_∞ controller synthesis by J-lossless coprime factorization. *SIAM Journal of Control and Optimization*, 30:522–547, 1992.

[36] M. Green, K. Glover, D. J. N. Limebeer and J. C. Doyle. A J-spectral factorization approach to H_∞ control. *SIAM Journal of Control and Optimization*, 28:1350–1371, 1990.

[37] M. Green and D. J. N. Limebeer. *Linear Robust Control.* Prentice-Hall, Englewood Cliffs, N.J., 1995.

[38] D. Hazony. Zero cancellation synthesis using impedance operators. *IRE Trans. on Circuit Theory*, 8:114–120, 1961.

[39] W. M. Hoddad and D. S. Bernstein. Robust stabilization with positive real uncertainty. *Systems and Control Letters*, 17:191–208, 1991.

[40] I. M. Horowitz. *Synthesis of Feedback Systems.* Academic Press, New York, 1963.

[41] I. M. Horowitz. A survey of quantitative feedback theory. *International Journal of Control*, 53:255–291, 1991.

[42] I. M. Horowitz and U. Shaked. Superiority of transfer function over state-variable methods in linear time-invariant feedback system design. *IEEE Trans. Automatic Control*, AC-20:84–97, 1975.

[43] S. Hosoe and T. Zhang. An elementary state space approach to RH^∞ optimal control. *Systems and Control Letters*, 11:369–380, 1988.

[44] T. Iwasaki and R. E. Skelton. All controllers for the general H^∞ control problem: LMI existence conditions and state space formulas. *Automatica*, 30:1307–1317, 1994.

[45] R. E. Kalman. Contributions to the theory of optimal control. *Bol. Soc. Mat. Mexicana*, 5:102–119, 1960.

[46] R. E. Kalman. On the general theory of control systems. In *Proceedings 1st IFAC World Congress, Moscow*, pages 481–491, 1960.

[47] R. E. Kalman. Mathematical description of linear dynamical systems. *SIAM Journal of Control*, 1:152–192, 1963.

[48] R. Kawatani and H. Kimura. Synthesis of reduced-order H^∞ controllers based on conjugation. *International Journal of Control*, 50:525–541, 1989.

[49] P. P. Khargonekar, I. R. Petersen and M. A. Rotea. H^∞ optimal control with state feedback. *IEEE Trans. Automatic Control*, 33:786–788, 1988.

[50] H. Kimura. Robust stabilizability for a class of transfer functions. *IEEE Trans. Automatic Control*, 27:788–793, 1984.

[51] H. Kimura. Generalized Schwarz form and lattice-ladder realizations of digital filters. *IEEE Trans. on Circuit and Systems*, 32:1130–1139, 1985.

[52] H. Kimura. Directional interpolation approach to H^∞ optimization and robust stabilization. *IEEE Trans. Automatic Control*, 32:1085–1093, 1987.

[53] H. Kimura. Directional interpolation in the state space. *Systems and Control Letters*, 10:317–324, 1988.

[54] H. Kimura. Conjugation, interpolation and model-matching in H^∞. *International Journal of Control*, 49:269–307, 1989.

[55] H. Kimura. State space approach to classical interpolation problem and its applications. In H. Nijmeijier and J. M. Schumacher (Eds.), *Three Decades of Mathematical System Theory*, pages 243–275. Springer Verlag, Berlin, 1989.

[56] H. Kimura. Application of classical interpolaton theory. In N. Nagai (Ed.), *Linear Circuits, Systems and Signal Processing*, pages 61–85. Marcel Dekker Inc., New York, 1990.

[57] H. Kimura. Linear fractional transformations and J-canonical factorization in H^∞ control theory. In *Proceedings of the American Control Conference*, pages 3085–3091, 1990.

[58] H. Kimura. (J, J')-lossless factorization based on conjugation. *Systems and Control Letters*, 19:95–109, 1992.

[59] H. Kimura. Chain-scattering representation, J-lossless factorization and H^∞ control. *Journal of Math. Systems, Estimation and Control*, 5:203–255, 1995.

[60] H. Kimura, Y. Lu and R. Kawatani. On the structure of H^∞ control systems and related extension. *IEEE Trans. Automatic Control*, 36:653–667, 1991.

[61] H. Kimura and F. Okunishi. Chain-scattering approach to control system design. In A. Isidori (Ed.), *Trends in Control; A European Perspective*, pages 151–171. Springer Verlag, Berlin, 1995.

[62] V. C. Klema and A. J. Laub. The singular value decomposition: Its computation and some applications. *IEEE Trans. Automatic Control*, 25:164–176, 1980.

[63] R. Kondo and S. Hara. On cancellation in H^∞ optimal controllers. *Systems and Control Letters*, 13:205–210, 1989.

[64] W. Kongprawechnon and H. Kimura. J-lossless conjugation and factorization for discrete-time systems. *International Journal of Control (to appear)*.

[65] V. Kucera. Algebraic riccati equation: Symmetric and definite solutions. In *Proceedings of the Workshop in the Riccati Equation in Control, Systems and Signals, Como*, pages 73–77, 1989.

[66] H. Kwakernaak. Minimax frequency domain performances and robustness optimization of linear feedback systems. *IEEE Trans. Automatic Control*, 30:994–1004, 1985.

[67] P. H. Lee, H. Kimura and Y. C. Soh. Characterisations of J-lossless time-varying systems via the chain-scattering approach. *Int. Journal of Robust and Nonlinear Control (to appear)*.

[68] P. H. Lee, H. Kimura and Y. C. Soh. (J, J')-lossless conjugations, (J, J')-lossless factorizations and chain-scattering approach to time-varying H^∞ control – one and two-block cases. *International Journal of Control (to appear).*

[69] D. J. N. Limebeer and B. D. O. Anderson. An interpolation theory approach to H_∞ controller degree bounds. *Linear Algebra and its Appl.*, 98:347–386, 1988.

[70] D. J. N. Limebeer, B. D. O. Anderson, P. P. Khargonekar and M. Green. A game theoretic approach to H^∞ control for time-varying systems. *SIAM Journal of Control and Optimization*, 30:262–283, 1992.

[71] D. J. N. Limebeer and G. Halikias. An analysis of pole zero cancellation in H^∞ control problems of the second kind. *SIAM Journal of Control and Optimization*, 25:1457–1493, 1987.

[72] D. J. N. Limebeer and G. Halikias. A controller degree bound for H^∞ optimal control problems of the second kind. *SIAM Journal of Control and Optimization*, 26:646–677, 1987.

[73] A. G. J. MacFarlane and I. Postlethwaite. Characteristic frequency functions and characteristic gain functions. *International Journal of Control*, 26:265–278, 1977.

[74] D. Q. Mayne. The design of linear multivariable systems. *Automatica*, 9:201–207, 1973.

[75] D. McFarlane and K. Glover. A loop shaping design procedure using H_∞ synthesis. *IEEE Trans. Automatic Control*, 37:759–769, 1992.

[76] G. Meisma. *Frequency Domain Methods in H^∞ Control*. Ph. D. dissertation submitted to University of Twente, 1993.

[77] R. Nevanlinna. Über beschränkte Funktionen, die in gegeben Punkten vorgeschriebene Werte annehmen. *Ann. Acad. Sci. Fenn.*, 13:27–43, 1919.

[78] G. Pick. Über die Beschränkungen analytischer Funktionen, welche durch vorgegebene Funktions werte bewirkt sind. *Mathematische Ann.*, 77:7–23, 1916.

[79] V. P. Potapov. The multiplicative structre of J-contractive matrix functions. *American Math. Soc. Translations*, 15:131–243, 1960.

[80] C. V. K. Prabhakara Rao and P. Dewilde. *System theory for lossless wave scattering*, Vol. 19 of *Operator Theory: Advances and Applications*. Birkhäuser, Boston, 1986.

[81] R. M. Redheffer. On a certain linear fractional transformation. *Journal of Mathematics and Physics*, 39:269–286, 1960.

[82] H. H. Rosenbrock. *State-Space and Multivariable Theory*. John Wiley, New York, 1970.

[83] H. H. Rosenbrock. *Computer Aided Control System Design*. Academic Press, New York, 1974.

[84] M. G. Safonov. *Stability and Robustness of Multivariable Feedback Systems*. MIT Press, Cambridge, 1980.

[85] M. G. Safonov, D. J. N. Limebeer and R. Y. Chiang. Simplifying the H^∞ theory via loop shifting, matrix pencil and descriptor concepts. *International Journal of Control*, 50:2467–2488, 1989.

[86] M. Sampei, T. Mita and M. Nagamichi. An algebraic approach to H^∞ output feedback control problems. *Systems and Control Letters*, 14:13–24, 1990.

[87] C. Scherer. H_∞ control by state feedback; An iterative algorithm and characterization of high gain occurrence. *Systems and Control Letters*, 12:383–391, 1989.

[88] J. Sefton and K. Glover. Pole/zero cancellations in the general H^∞ problem with reference to a two block design. *Systems and Control Letters*, 14(3), 1990.

[89] U. Shaked and I. Yaesh. A simple method for deriving J-spectral factors. *IEEE Trans. Automatic Control*, 37:891–895, 1992.

[90] L. C. Siegel. *Symplectic Geometry*. Academic Press, 1964.

[91] G. Stein and J. C. Doyle. Singular values and feedback: Design examples. In *Proceedings of the Allerton Conference. Urbana*, 1978.

[92] W. Sun, P. P. Khargonekar and D. Shim. Solution to the positive real control problem for linear time-invariant systems. *IEEE Trans. Automatic Control*, 39:2034–2046, 1994.

[93] G. Tadmor. Worst-case design in the time domain: The maximum principle and the standard. *Mathematics of Control, Signals and Systems*, 3:301–324, 1990.

[94] J. G. Truxal. *Control System Synthesis.* McGraw-Hill, New York, 1955.

[95] M. C. Tsai and I. Postlethwait. On J-lossless coprime factorization approach to H^∞ control. *International Journal of Robust and Nonlinear Control*, 1:47–68, 1991.

[96] M. C. Tsai and C. S. Tsai. A chain scattering matrix description approach to H^∞ control. *IEEE Trans. Automatic Control*, 38:1416–1421, 1993.

[97] M. S. Verma and E. A. Jonckheere. L^∞ compensation with mixed sensitivity as a broadband matching problem. *Systems and Control Letters*, 4:125–130, 1984.

[98] M. Vidyasagar and H. Kimura. Robust controllers for uncertain linear multivariable systems. *Automatica*, 22:85–94, 1986.

[99] J. C. Willems. From time series to linear system–part 1. Finite dimensional linear time invariant systems. *Automatica*, 22:561–580, 1986.

[100] J. C. Willems. Paradigms and puzzles in the theory of dynamical systems. *IEEE Trans. Automatic Control*, 36:259–294, 1991.

[101] K. H. Wimmer. Monotonicity of maximal solutions of algebraic Riccati equations. *Systems and Control Letters*, 5:317–319, 1985.

[102] X. Xin and H. Kimura. (J, J')-lossless factorization for descriptor systems. *Linear Algebra and its Appl.*, 205-206:1289–1318, 1994.

[103] X. Xin and H. Kimura. Singular (J, J')-lossless factorization for strictly proper functions. *International Journal of Control*, 59:1383–1400, 1994.

[104] D. C. Youla. A new theory of cascade synthesis. *IRE Trans. on Circuit Theory*, 9:244–260, 1961.

[105] D. C. Youla and M. Saito. Interpolation with positive-real functions. *Journal of The Franklin Institute*, Vol. 284:77–108, 1967.

[106] G. Zames. On the input-output stability of time-varying nonlinear feedback systems, part I and part II. *IEEE Trans. Automatic Control*, 11:228–238, 465–476, 1966.

[107] G. Zames. Feedback and optimal sensitivity: Model reference transformations, multiplicative seminorms, and approximate inverses. *IEEE Trans. Automatic Control*, 26:301–320, 1981.

[108] G. Zames and B. A. Francis. Feedback, minimax sensitivity, and optimal robustness. *IEEE Trans. Automatic Control*, 28:585–601, 1983.

[109] K. Zhou. On the parametrization of H^∞ controllers. *IEEE Trans. Automatic Control*, 37:1442–1446, 1992.

Printed in the United States
By Bookmasters